WATER RESOURCES PLANNING, DEVELOPMENT AND MANAGEMENT

Edited by **Ralph A. Wurbs**

Water Resources Planning, Development and Management
http://dx.doi.org/10.5772/45847
Edited by Ralph A. Wurbs

Contributors

Makarius Victor Mdemu, Sandra Mesquita, Rachel Noble, Nikki Funke, Marius Claassen, Shanna Nienaber, Arturo Cruz-Falcón, James Tolan, Ralph Wurbs, Eric Koomen, Gregory Breetzke, William Critchley, Jianli Liu, Pengzhi Lin, Changbo Jiang

Published by InTech

Janeza Trdine 9, 51000 Rijeka, Croatia

Publishing Process Manager Dejan Grgur
Technical Editor InTech DTP team
Cover InTech Design team

Additional hard copies can be obtained from orders@intechopen.com

Water Resources Planning, Development and Management, Edited by Ralph A. Wurbs
p. cm.
ISBN 978-953-51-1092-7

For EU product safety concerns:
IN TECH d.o.o., Prolaz Marije Krucifikse Kozulić 3, 51000 Rijeka, Croatia,
info@intechopen.com or visit our website at intechopen.com.

Meet the editor

Ralph Wurbs holds the Arthur McFarland endowed professorship in the Zachry Department of Civil Engineering and is also a Professor in the interdepartmental Graduate Program in Water Management and Hydrologic Sciences at Texas A&M University. He is a Fellow of the American Society of Civil Engineers and Diplomate of the American Academy of Water Resources Engineers. He holds B.S., M.S., and Ph.D. degrees from Texas A&M University, University of Texas at Arlington, and Colorado State University.

Contents

Preface

Effective planning, development, and management of water resources, locally, regionally, and globally, are crucial for human welfare, economic prosperity, and environmental protection. Water is essential to all of us. Human health and socioeconomic welfare is dependent on adequate supplies of suitable quality water. The vitality of natural ecological systems is dependent on mankind's stewardship of water resources. Water is diverted from rivers, lakes, and aquifers to supply municipal, domestic, livestock, agricultural irrigation, energy production, and various industrial uses. Conversely, too much water results in socioeconomic damages and loss of life due to flooding. Flood risk mitigation, drainage, and erosion control are important concerns along with water supply, health and sanitation, and water quality protection.

Water has played a key role in the development of civilizations since ancient times. Problems, needs, and opportunities have grown dramatically during the past century with population growth, urban and industrial development, expansion of irrigated agriculture, aging infrastructure, pollution and environmental degradation, and technological advances. Continually growing demands on limited water resources will intensify future challenges.

This book addresses a broad range of fundamental issues that are essential to effective planning, development, and management of water resources. The authors represent diverse fields of science, engineering, and water management as well as diverse geographical perspectives. Although applications in particular countries or regions are discussed in the individual chapters, the concepts and methods presented are generally applicable throughout the world. The book provides authoritative coverage of an array of innovative state-of-the-art methods for supporting important decisions faced by water professionals and society.

Chapter 1 describes concepts of developing scenarios in dealing with uncertainties in water planning. Chapter 2 is a state-of-the-art assessment of capabilities for monitoring microbiological indicators of water quality. Chapter 3 provides a comparative assessment of the soil loss estimator method applied in South Africa for predicting erosion and the widely applied universal soil loss equation. The biological response to freshwater inflows to estuaries is explored in Chapter 4. The concept of constructing freshwater reservoirs in the seawater near river mouths to capture river flows is explained in Chapter 5. Chapter 6 describes methods for assessing the productivity of water allocation schemes for large scale rice irrigation. Chapter 7 investigates factors controlling groundwater recharge. Mechanisms for allocating limited water resources between numerous water users are described in Chapter 8.

The concepts and methods presented in this book should prove beneficial to both water management practitioners and research scientists and engineers. As editor, I sincerely thank the authors for their contributions to the important interdisciplinary fields that advance water planning, development, and management for the benefit of people throughout the world.

Ralph A. Wurbs, Ph.D., P.E., D.WRE
Arthur McFarland Professor of Civil Engineering
Professor of Water Management and Hydrologic Science
Texas A&M University
College Station, Texas, USA

Chapter 1

Development and Uptake of Scenarios to Support Water Resources Planning, Development and Management – Examples from South Africa

Nikki Funke, Marius Claassen and Shanna Nienaber

Additional information is available at the end of the chapter

http://dx.doi.org/10.5772/52577

1. Introduction

The international agenda on water resources development reflects societal needs, political agendas, economic realities and the state of resources. The industrial revolution, which started in the 18th century, brought social and economic prosperity but also marked a major shift in humanity's impact on the earth's systems. This shift is now referred to as the Anthropocene [1], where humans have brought such vast and unprecedented changes to the planet that this era represents a new geological time interval [2]. Societal needs have shifted since the 1940s from a need for modest food production to a need for increased agricultural productivity that has been met by high yield crops, the use of pesticides, the application of fertiliser and advanced agricultural techniques. This development has averted food shortages, but has also resulted in humanity having to pay a heavy price in terms of increased water use and energy consumption, as well as environmental degradation [3].

From the early 1970s a series of events and key documents has promoted an integrated approach to sustainable development. The 1972 United Nations Conference on the Human Environment considered the need for a common outlook towards the preservation and enhancement of the human environment [4]. The World Commission on Environment and Development advanced this agenda in their report 'Our Common Future', with an emphasis on sustainable development promoting harmony among human beings and between humanity and nature [5]. The International Conference on Water and the Environment that took place in Dublin in 1992 resulted in the development of four guiding principles [6]. These principles, commonly referred to as the Dublin principles, state that: water is a finite resource with economic value and social implications; local communities must participate in water management; water resources management must be developed within a set of policies; and the role of rural populations and women should be recognised. This led to the Rio Declaration and the adoption of Agenda 21, which is a comprehensive plan of action to be implemented globally, nationally and locally in every area in which humanity impacts on the environment [7]. This declaration subsequently became the blueprint for sustainable development world-wide [8].

Uncertainties about societal, economic, political and environmental aspects have proved to be a considerable obstacle to the implementation of sustainable development. Here follow a few examples of such uncertainties. In 1980, the World Development Report of the United Nations [9] estimated that the world population would reach 6.029bn by the year 2000. Five years later, the estimate was updated to 6.088bn [10], with further updates at five yearly increments resulting in estimates of 6.194bn and 6.123bn [11, 12]. The actual population in the year 2000 turned out to be 6.188bn [13]. Future economic development is also uncertain, with the annual growth in Figure 1 showing how the world average varies significantly between years and also how the growth of individual countries (South Africa in this case), does not necessarily follow the global trend and is even more variable between years.

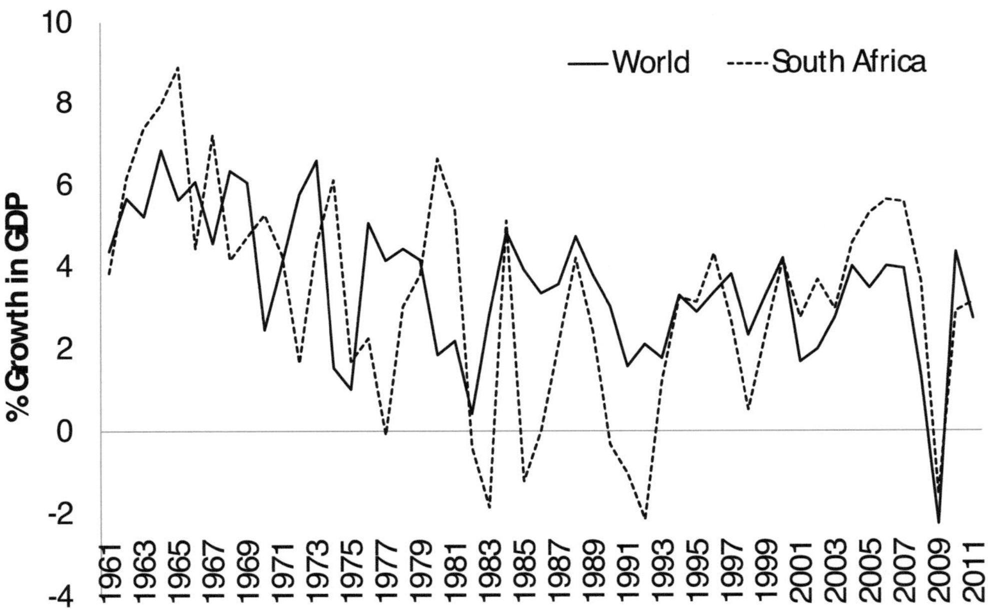

Figure 1. Annual economic growth between 1961 and 2011 [13]

Environmental conditions also vary significantly over time and space, with Figure 2 illustrating the annual deviation of rainfall over southern Africa. This uncertainty is exacerbated by climate projections, which suggest that freshwater resources are vulnerable and have the potential to be strongly impacted by climate change, with wide-ranging consequences for human societies and ecosystems [14].

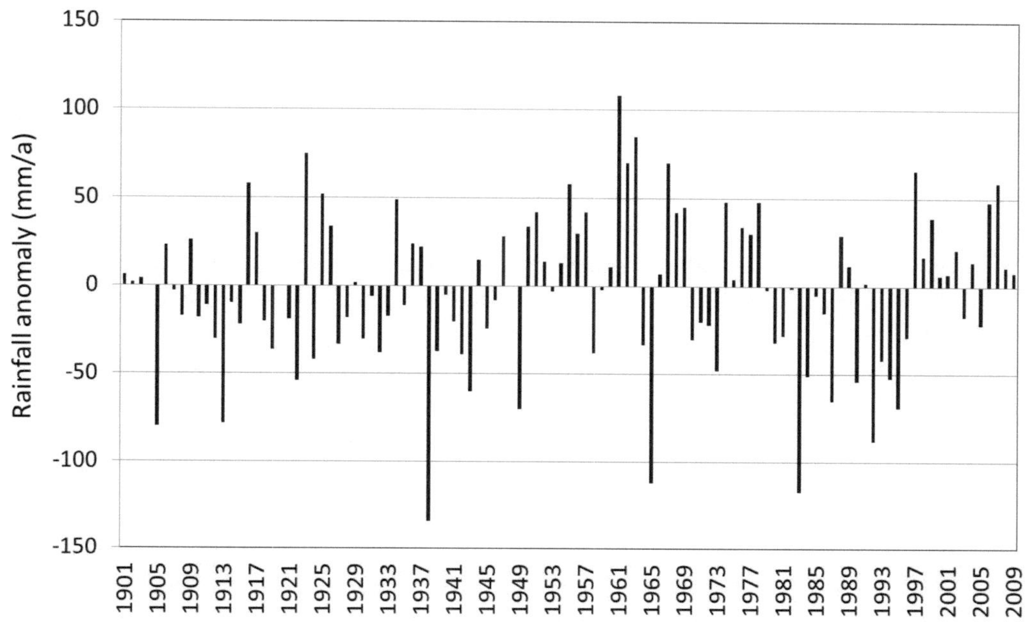

Figure 2. Annual rainfall anomalies for the southern African region (1901-2009; Adapted from [15]).

A question that emanates from the realisation that we live in a changing world where change is unpredictable is, 'How do we plan for the future?'

Water use in South Africa was first regulated through the Irrigation and Conservation of Waters Act (Act No. 8 of 1912), which managed the use of water from public streams for domestic, irrigation and industrial purposes [16]. The Water Act (Act No. 54 of 1956) further regulated water use by providing for the control of water pollution and the more effective protection of water resources. The variable distribution of water required the development of infrastructure to capture, store and distribute water. The subsequent expansion of mines, industries and urban areas created a demand for further infrastructure development. When this demand further increased and the social and economic issues in South Africa became increasingly complex in the 1990s as the country was transitioning from *apartheid* to democracy, a shift in thinking was required. As it became clear that engineering solutions to increase water supply were not sustainable, a holistic strategy to meet future needs became more popular [17].

The new National Water Act (Act No. 36 of 1998) [18] emphasised water resources management at national and catchment scales, made specific provisions for the protection of water resources, established mechanisms to ensure equitable and efficient water use and promoted participatory management. The National Water Resources Strategy [19] addressed the balance of future water supply and demand by establishing scenarios. The demand scenarios were based on population growth by 2025, with the high population scenario at 54 million people and the low population scenario at 50 million people. It also established economic growth scenarios, with the upper scenario assuming 4% growth in GDP and the less favourable scenario assuming 1.5% [19].

While there has been much progress in water infrastructure development for services (public benefit), the backlog in issuing water use licenses (mostly for private benefit) stood at 4 318 in 2011 [20]. The protection of water resources has suffered as a result of the government's drive to achieve social and economic development, with South Africa ranked 128 out of 132 countries in the Environmental Performance Index [21]. The National Water Act provides for a balance of responsibilities, ranging from the Minister and Director General at the national level, to Catchment Management Agencies (CMAs) at the basin level and Water User Associations (WUAs) at a sub-basin level. Progress has been slow as after 14 years after the promulgation of the Water Act, only two CMAs (out of the 19 intended) have been established [20]. It can be argued that many hurdles have to be overcome to fully realise cooperative governance for Integrated Water Resources Management (IWRM), with inadequate human and institutional capacity being one of the main factors limiting the efficient management of water resources in South Africa [22]. To illustrate this point: the country's Department of Water Affairs (DWA) reported having 4 286 people in its employment in September 2010, while 1 155 posts were vacant at the time [20].

From the discussion above it becomes clear that we live in a world with social, economic and environmental conditions that are variable and difficult to predict, and the water sector is no exception. This uncertainty provides a challenging environment for policy and institutional development. Scenarios are one way of attempting to achieve a desired outcome in an uncertain and variable future [23]. The rest of this chapter will examine the research question, 'How are scenarios able to achieve impact in an uncertain world, with a particular focus on water resources planning, development and management?' The body of this chapter focuses on the research method, presents an overview of scenario development and the importance of scenario development and how they facilitate more effective water resources planning, development and management, focuses on a few select South African scenarios and the impact they have had and then turns to discussing the impact of scenarios in general. The conclusion wraps up the learning from this chapter and suggests a way forward in terms of future research and designing scenarios for impact.

2. The ability of scenarios to achieve impact in an uncertain world with a focus on water planning, development and management

2.1. Method

The authors of this chapter conducted an exploratory study on the ability of scenarios to achieve impact in an uncertain world, with particular reference to water planning, development and management. They conducted a review of scenario planning literature in the water and other sectors, and also considered literature focusing specifically on the impact of scenarios. The authors also considered literature on the impact of scientific research and on the science-policy interface. This was accompanied by a search of major databases (e.g. Google Scholar, EBSCO Host and Scopus) to determine where and how the four scenarios discussed in this chapter have been cited. In addition, the authors interviewed selected stakeholders in the water and other sectors who are likely to have been exposed to scenarios and who may use scenarios when making decisions in their workplace.

2.2. Scenarios and their importance in the water sector

2.2.1. The history of scenario development

The concept of scenario planning has its origin in military applications, with the US Air Force developing 'scenarios' of what the enemy might do and preparing alternative strategies. It was thus aimed at achieving a desired outcome in an uncertain future [24]. At the end of the 1940s, researchers at the RAND Corporation started to investigate the scientific use of expert opinion in planning for the future [25]. The Royal Dutch Shell company employed scenario tools to good effect in the 1970s, when they improved their size and profitability by being prepared to act quickly during the oil price shock of 1973 [26]. In summarising definitions of scenarios, scenarios can be described as a narrative description of a possible state of affairs or development over time, that they are useful to communicate speculations about the future to promote discussion and feedback, and that they can dramatise trends and alternatives, explore the impacts and implications of decisions, choices and policies, and provide cause-and-effect explanations [24].

Clem Sunter is credited with popularising the use of scenarios in South Africa, with 'The World and South Africa in the 1990s', which describe the 'High Road' and 'Low Road' scenarios [27]. The publication was based on work from Anglo American Corporation teams in London and Johannesburg. Subsequently, Adam Kahane facilitated a process that became known as the Mont Fleur scenario project, which was launched in 1992. It explored the question of 'What will South Africa be like in the year 2002?' These scenarios were arrived at collaboratively by a very broad group [28]. The Department of Arts, Culture, Science and Technology (DACST) also deployed scenarios and technology foresighting in the development of South Africa's National Research and Development Strategy, with Kahn initiating and leading the development of the South African National Research and Technology Foresight Project [29]. The Dinokeng team [30] developed '3 Futures for South Africa', which characterised future scenarios based on the effectiveness of the state and the engagement of society. Some of the recent scenario projects in the water sector include the World Business Council for Sustainable Development report on 'Business in the World of Water: WBCSD Water Scenarios to 2025' [31], and the Global Research Alliance (GRA) report on 'Science and Technology-based Water Scenarios for sub-Saharan Africa' [32].

2.2.2. *The importance of the use of scenarios in water resources planning, development and management*

Scenarios are important and useful to water resources planning, development and management in a number of ways. In the South Africa context, in particular, scenario development processes have been instrumental in initiating strategic conversations among scenario workshop In the South African context, (e.g. the transition from *apartheid* to democracy), and have helped develop a common language among people with widely divergent views [28]. Those involved in scenario development processes may be inspired to think more broadly about the future and the forces creating it. They may also realise how their particular actions may help to create a desired future [33]; and they may have suggestions about which options exist to direct target audiences on to a desirable path [28]. The knowledge that scenarios generate can therefore potentially empower role players in the water sector and other sectors to engage in participative governance by equipping them with insights into potential futures they may face, and making them aware of the implications of certain decisions, behaviours and actions [23]. Finally, the advantage of communicating scenarios as stories is that they have the psychological impact that other more academic means of communication, for example, graphs and equations, lack. Stories give order and meaning to events, which is crucial for imagining future possibilities [34].

2.3. Some South African scenarios: Overview and impact

The discussion in this chapter and the research question were inspired by the development of the Water Sector Institutional Landscape by 2025 scenarios. These scenarios were the main output of a research project led by the authors. In particular, the authors are interested in how these scenarios could be used by potential end-users. Given this question and the importance and potential usefulness of scenarios in facilitating decision-making in a context of uncertainty, it becomes important to reflect on some examples of scenarios that have been developed in South Africa at different points in history and to learn from the impact they have had on different sectors, including the water sector. These scenarios are discussed in chronological order. The section starts with the High Road/Low Road scenarios that were developed late in the *apartheid* era and on the cusp of South Africa's transition to democracy. Secondly, the Mont Fleur scenarios, which were developed during the democratic negotiations, are discussed. Thirdly, the section focuses on the Dinokeng scenarios that were developed in 2009, the year a new president came to power and a serious economic crisis shook the world. The section concludes with the Water Sector Institutional Landscape scenarios that focus on potential futures of the South African water sector in 2025.

2.3.1. *High road/Low road*

2.3.1.1. *Overview and process*

The High Road/Low Road scenarios were an initiative by the Anglo American Corporation in the early 1980s and aimed to look into some less conventional approaches to business planning and future investment decisions, given the international economic turbulence of the 1970s

http://dx.doi.org/10.5772/52577

and the resultant slump in commodity markets. During this time, South Africa's economic performance was poor and several events resulted in the country becoming increasingly isolated and the government resorting to a rule of force. Careful and gradual reforms by the *apartheid* government in the middle to late 1980s and increasing attempts by members of the white establishment to reach out to black leaders in exile, led to the eventual unbanning of the African National Congress (ANC) and the release of Nelson Mandela in 1990 [35].

The scenarios involved a large-scale exercise with numerous contributors, notably Pierre Wack and Ted Newland, as well as Clem Sunter. Most of the effort went into developing global scenarios which were based on the analysis of key 'drivers' (for example, demography, technology and societal values) of developments in Japan, the USA and USSR (then regarded as the main players of the world economy), and also the ingredients for success of 'winning' nations and world class companies. This work then provided the basis for the South African scenarios. In essence, these scenarios focus on the choice the country was facing to either (through consultation and negotiation) travel on the 'High Road' to a non-racial democracy and increasing prosperity, or, to continue on the 'Low Road' of confrontation, conflict and falling incomes (as a repressive, centralised society and controlled economy) and ending up as a 'waste land' [35].

The scenarios conclude with the need for a 'common vision' to help launch South Africa into the more desirable 'High Road' scenario. This common vision entails putting South Africans first (looking beyond different races and groups), to turn the country into a 'winning' nation and to work towards achieving a certain income per head, all of which would be reached through negotiation [28].

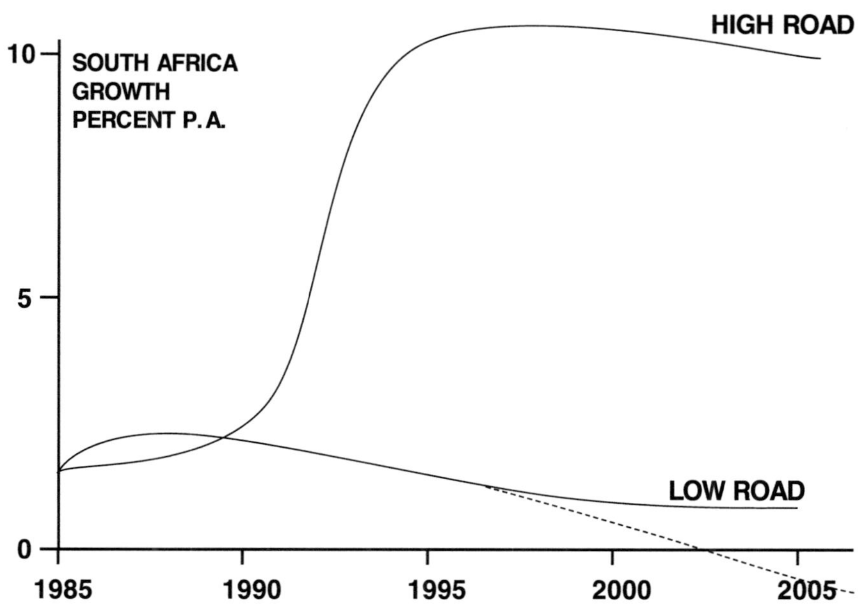

Figure 3. The High Road/Low Road scenarios depicting two possible future trajectories for South Africa [27]

2.3.1.2. Dissemination and impact

Within a year, starting in 1986, Clem Sunter presented the 'High Road/Low Road' scenarios to 230 (mostly white) audiences at various levels of society, thereby reaching between 25 000 to 30 000 people [35]. Senior politicians of the ANC were also one of Sunter's audiences before the eventual negotiated settlement was reached [28]. The message of the scenarios seems to have made a big impression on the audiences as it was ultimately positive and encouraged people in the country to take it into their own hands to get on to the 'High Road', without being prescriptive about how this should be done [35]. In particular, the High Road/Low Road scenarios also seem to have contributed somewhat to the shift in thinking in government circles, and indeed as supporting evidence for a need for change, which eventually brought about a political transition. In conclusion then, the High Road/Low Road scenarios started out as a corporate scenario project and resulted in a brilliant communication exercise, both in terms of content and style of presentation, that reached thousands beyond the initial intended audience and paved the way for more prominent South Africa scenario exercises to come [35].

In terms of uptake in the scientific and decision-making community, Clem Sunter's book 'South Africa and the World in the 1990s' has been widely cited and includes discussions of a range of topics. These include reflections on various elements of the political and economic transformation of South Africa, the future of Africa, scenario development and planning and globalisation. The citations include a variety of different sources, including books, journal articles, theses and reports. These sources are mostly from the economic, management and social sciences, but also from the health and environmental sciences.

While no examples could be found of the use of the High Road/Low Road scenarios in the water sector, it is likely, judging from the fact that Sunter presented these scenarios to such a wide range of audiences, that some members of government and other stakeholders in the water sector would have been exposed to them in the late 1980s or early 1990s. South Africa's new water legislation certainly reflects the thinking associated with the High Road scenario, with emphasis on introducing ground-breaking new principles into the governance of South Africa's water resources. Though somewhat outdated now, the High Road/Low Road scenarios serve as a reminder of where South Africa could be headed at any point in history. In terms of water resources, South Africa is in need of thoughtful planning, development and management if its water resources are to continue to meet the needs of its ever growing and developing population.

2.3.2. Mont Fleur

2.3.2.1. Overview and process

The Mont Fleur scenarios were developed in South Africa between 1990 and 1994. Key events during this time were the release of Nelson Mandela, and the legalisation of the

ANC, Pan African Congress (PAC) and South African Communist Party (SACP) [36]. The country's first racially inclusive elections were also held at this time. Given this political climate, multiple forums emerged that brought a broad range of stakeholders together to try to develop a new way forward for South Africa. In particular, issues such as housing, education, and constitutional reform received attention [35, 36].

The Mont Fleur scenarios formed a part of this process and essentially tried to encourage debate, thinking and imaginative ideas around how to shape the first ten years of the 'new' South Africa and also to illustrate how certain choices would steer the country towards different outcomes. The Mont Fleur scenario team was made up of a diverse group of 22 prominent South Africans, including politicians, activists, academics and business people [36].

The Ostrich scenario represents a continuation of the *status quo* in South Africa and suggests that no negotiated settlement would be reached and that government would continue to be non-representative [37].

Figure 4. The Ostrich scenario [36]

The Lame Duck scenario suggests a South Africa where a settlement would have been achieved but where the transition to a new dispensation would be slow and indecisive [37].

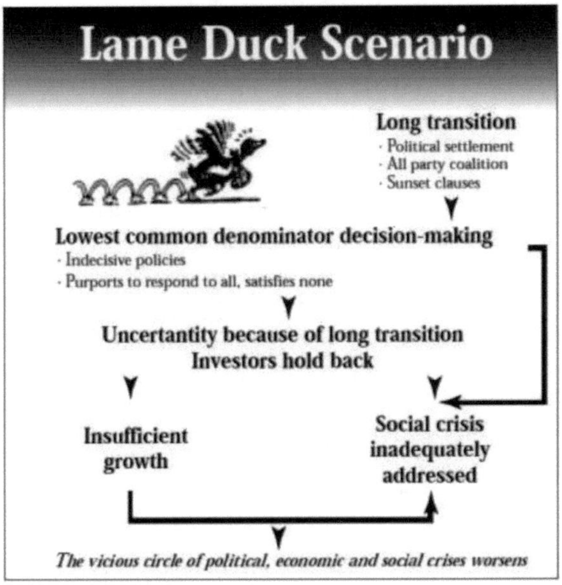

Figure 5. The Lame Duck scenario [36]

The Icarus scenario suggests a rapid transition to a new government that would push for populist and unsustainable economic policies [37].

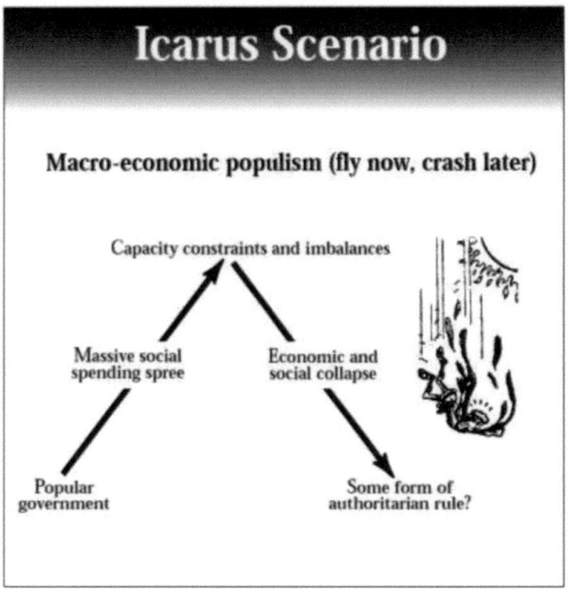

Figure 6. The Icarus scenario [36]

The Flight of the Flamingos scenario depicts a government that would choose sustainable policies that would lead the country towards inclusive growth and a maturing democracy [35, 36, 37].

Figure 7. The Flight of the Flamingos scenario [36]

By means of a process of negotiation and reflection on different drivers and concerns, the Mont Fleur scenario team was able to articulate a range of potential outcomes for South Africa during the 1992 to 2002 period. This also helped to clarify the goals and aspirations related to where the country should be heading.

2.3.2.2. Dissemination and impact

A variety of dissemination techniques were used by the Mont Fleur team. Key to this process was the fact that each of the individual participants took responsibility for spreading the message of these scenarios. They did this by presenting and discussing the scenarios with more than 50 different groups of people including political parties, companies, academics, trade unions and civil society organisations [36]. This was possible given the diverse background that the team came from. Over and above this process, the scenarios were condensed into an easily accessible 14 page document. This document was distributed to national newspapers. A short video was also produced that combined cartoons with presentations by team members [35, 36].

The impacts of this scenario development process are subtle. Three key points are important in this regard. Firstly, Mont Fleur, along with other processes taking place in South Africa at the time, helped to establish a common language and understanding about the challenges facing the country and the way forward. This was because participants focused on an issue of common concern for all: 'the future of South Africa'. Secondly, although participants could not agree on one major solution to South Africa's problems, they could agree that certain solutions would not work (such as armed revolutions, continued minority rule and socialism). Thirdly, through an informal process of open conversation, participants who had not expected to agree with each other found common ground and shared understandings about the future of the country [36]. Given these points it is clear that the impact that the Mont Fleur scenarios had was first and foremost on the individuals who participated in the process. There was subsequently a more indirect impact on broader society once these individuals started presenting the scenarios to their various constituencies. Given the widely publicised nature of South Africa's political transition, these scenarios also gained popularity overseas [35].

The Mont Fleur scenarios have also been cited in a range of publications. These citations occur in journals, books, conference papers, dissertations and magazines that focus on a range of different disciplines, namely the social, natural and technical sciences. Given this broad interest, the publications cover a broad range of topics most of which are geared towards futures research, democratic transition and strategic planning. This citation record illustrates that the Mont Fleur scenarios seem to have had a considerable impact on the academic community.

Whilst the Mont Fleur scenarios are not obviously related to the South Africa water sector, they did contribute to setting a precedent for using scenario development for planning purposes in South Africa. So, for instance, as mentioned above, the National Water Resources Strategy established a set of water demand scenarios. As with the High Road/Low Road scenarios, the Mont Fleur scenarios were part of the thinking and move towards democratic transition in South Africa. As a result of and in order to complement this change, the water sector was fundamentally transformed and restructured.

2.3.3. Dinokeng scenarios

2.3.3.1. Overview and process

The Dinokeng scenario team consisted of 35 leaders from civil society, government, business, political parties, public administration, trade unions, religious groups, academia and the media. The scenario development process was sponsored by the financial institutions Old Mutual and Nedbank who believed that, 15 years into South Africa's democracy, it was important to initiate a reflective and constructive debate about the country's future. According to the Dinokeng scenario team, some of the most prominent challenges facing South Africa are unemployment and poverty, safety and security, education and health. These challenges appear all the more grave in the context of a volatile global economic market, and a global economic crisis that shook the world when these scenarios were developed in 2009 [30].

The Dinokeng scenario team agreed that South Africa needs to realise that the country has failed to appreciate or understand the imperatives of running a modern democratic state, and that there is a problem with the country's self-interested, unethical and unaccountable leadership across all sectors. Additional problems include a weak state that is increasingly less capable of addressing the country's critical challenges, and a population that is either not interested and is showing a growing dependence on the state to provide for everything, or has become co-opted into government or party structures since 1994 [30].

The scenario team developed three possible scenarios which the country could be heading into:

Firstly, the Walk Apart scenario suggests the state becoming increasingly weak and ineffective, and the population, which is looking out for its own interests, eventually losing patience with the state and resorting to protest and unrest to make its views heard. Because the state is unable to meet the population's demands and expectations, it responds brutally, and the result is a spiral of resistance and repression. The Walk Apart scenario therefore suggests a need for South Africans to address their critical challenges, to build state capacity and to organise themselves to engage government in a constructive way, in order to prevent themselves from heading towards disintegration and decline [30].

Secondly, the Walk Behind scenario suggests the state becoming increasingly confident and strong in terms of leading and directing development, fuelled by the fact that civil society is becoming more and more dependent and compliant. The problem is that the state does not have the capacity to address the critical challenges the country is facing on its own. The message of this scenario is that state-led development cannot be successful if there is insufficient state capacity. Furthermore, if the state intervenes constantly and dominates all other sectors, it will crowd out private business and civil society initiatives and will end up creating a population that is complacent and dependent on the state [30].

Thirdly, the Walk Together scenario suggests the state becoming collaborative and increasingly listening to its citizens and leaders from different sectors, engaging with critical voices, and consulting and sharing authority in order to work towards long-term sustainability. In this scenario there is also a focus on a population that takes leadership and holds government accountable and shows an active interest in policy development and outcomes. It is important that South Africans re-engage, that the capacity of the state is strengthened and that leaders from all sectors think beyond their own self-interest and contribute to nation-building [30].

In conclusion, the present contains the seeds for all three futures to be realised. For a healthy democracy and strong socio-economic development to persist, it is important to have a healthy interface between an effective state and an alert and involved population; the nature of this interface is likely to determine the future of the country [30].

2.3.3.2. Dissemination and impact

In terms of dissemination and impact, once the Dinokeng scenarios on possible futures for South Africa had been developed, the messages of these scenarios were shared with a range of stakeholders. This engagement was followed up with a media and engagement campaign to communicate the Dinokeng scenarios to a variety of organisations, groups and communities across South Africa [30]. The Dinokeng scenarios and the process around their development were also placed on the Dinokeng scenarios website, which is a user-friendly resource for those who are interested in finding out more about these scenarios. The Dinokeng scenarios text is also available for download here.

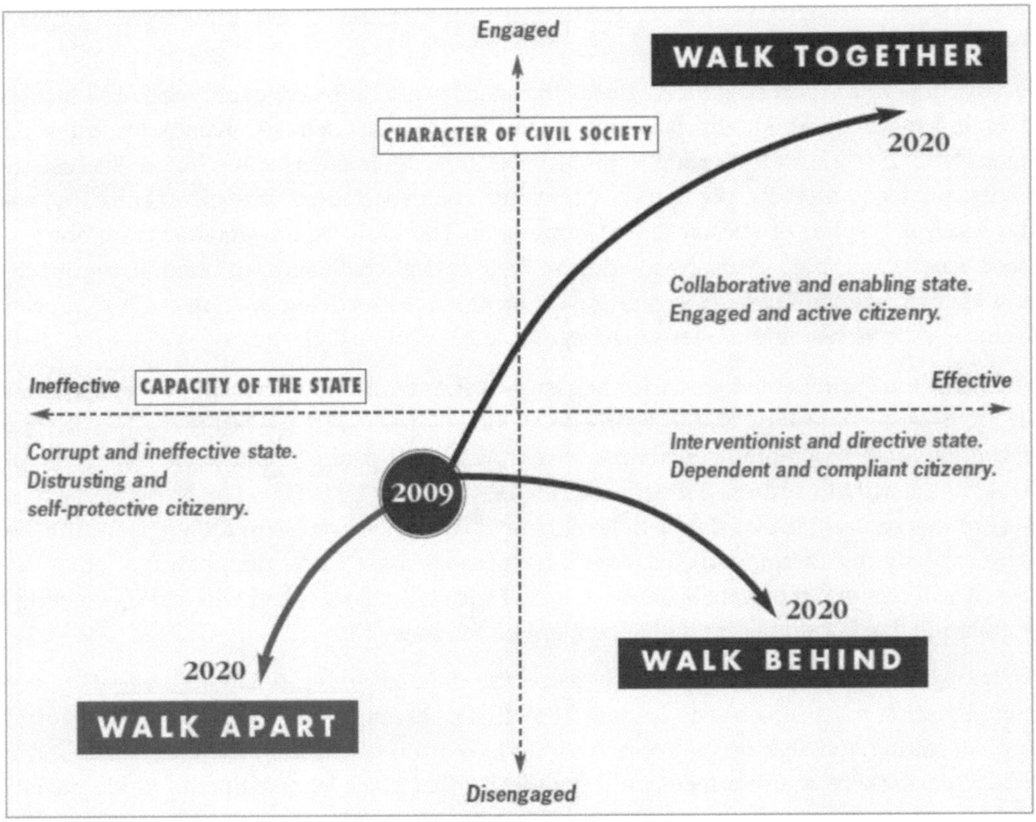

Figure 8. The Dinokeng scenarios [30]

A database search showed that these scenarios have been cited in a wide range of publications. These publications include discussion papers, theses, conference presentations, books and journal papers. The topics of the publications that cited the Dinokeng scenarios are wide-ranging and include issues around interrogating and addressing social issues related to South Africa's democracy, such as local government, education, housing, poverty, unemployment and food security. Many of these topics have a future-centred focus, e.g. investigating South Africans' perceptions about the future, or planning for the future in local government structures. The fact that the Dinokeng scenarios were cited in different kinds of publications and across different subject matters indicates that, at least among the research community, the scenarios were widely distributed and taken up by researchers from different social science-based backgrounds and interests.

A question that arises here is to what extent the Dinokeng scenarios may be of relevance to water resources planning, development and management. While no examples of their use in the water sector were found, it can be argued that the insights provided by these scenarios would prove valuable in focusing on resolving some of the water governance related issues South Africa is currently facing. Examples include problems around water pollution resulting from ineffective waste water treatment and mine and industrial effluents, and water service delivery to previously disadvantaged communities. Those who need to address these water governance related problems could benefit from taking into account the need for maintaining a balance between strong and effective leadership in all sectors and an interested and engaged population, and reflecting on the different future directions such a relationship or lack thereof could take.

2.3.4. Water sector institutional landscape by 2025

2.3.4.1. Overview and process

An example of scenario development with particular reference to the South African water sector is the South African Water Research Commission's (WRC) Water Sector Institutional Landscape by 2025 scenarios, developed by the authors in 2011 with the assistance of Chan tell Illbury as facilitator, and in consultation with a range of water sector related experts and stakeholders. The focus of the scenario development was on water resources management in South Africa, also with relevance to the water services sector. The aim of these scenarios was to build knowledge about key drivers and uncertainties that relate to the future of the South African water sector, and specifically about the context in which water institutions may operate in future [23].

The knowledge for this project was generated through a structured research process to target existing and new institutional structures and to ensure the involvement and participation of a broad range of stakeholders. The aim of this engagement was to identify water-related needs, priorities and uncertainties based on a wide range of perspectives. A broad range of methods was employed to include stakeholders from both rural and urban environments and with different cultures and educational backgrounds. These included interactive workshops, semi-structured interviews, and a web-based survey. This process was characterised by continuous assessment, learning and adaptation [23].

The key drivers and uncertainties that were identified were subsequently translated into different scenarios that hold potential implications for social and economic development, as well as water resources and services in South Africa. The four scenarios were derived from a matrix with two axes that represent the ability of the decision-making paradigm of water institutions to deal with 'complexity' (refer to the x-axis of the diagram), and the reconciliation of environmental, social and economic demands of present and future generations (referred to as 'sustainability' on the vertical or y-axis of the diagram) [23].

Four possible scenarios emerged from the matrix. The Greedy Jackal scenario depicts a South Africa where water is scarce but government still struggles to meet developmental demands and address backlogs. Under these urgent socio-economic circumstances, environmental responsibility is not prioritised. Despite this the need for a multidisciplinary response to complex water challenges is acknowledged [38].

The Wise Tortoise scenario suggests that a paradigm shift has occurred resulting in a water sector that is multi-layered and engages many different sectors given the strategic importance of the resource in all facets of development. This approach allows for proactive management rather than crisis response to challenges [38].

The Busy Bee scenario suggests that the water sector is defined by great intentions but does not follow up on these with necessary actions. Thus, whilst rhetoric embraces sustainability, in practice there is limited economic and social development to support this process. Part of the challenge is a lack of civil society engagement and failure to embrace the complexity facing water resources management [38].

The Ignorant Ostrich scenario suggests that government fails to recognise water as central to development. As such they rush to implement politically appealing but imbalanced and short term solutions. Civil society is not engaged in decision-making and the complexity inherent to the water sector is overlooked [38].

Figure 9. The Water sector institutional landscape by 2025 scenarios [38]

2.3.4.1. Dissemination and impact

The scenarios were printed by the WRC in the form of a colourful booklet and subsequently have been disseminated to some stakeholders. The scenario document and technical report documenting the scenario development process are also available online. While much more could have be done in terms of dissemination, this was not a component required by the project's funders and was therefore not planned into the project process from the start of the project. Therefore no funding was available to carry out this important part of the scenario development process. Nonetheless, these scenarios have the potential of feeding into the decision-making processes of water resources managers and decision-makers, but could also potentially empower a range of other role players in the water sector to engage in participative governance [23].

By studying the dissemination, impact and lessons learned from the South African scenarios discussed above, along with other literature related to the impact of scientific research and the science-policy interface, it is possible to distil some lessons and challenges relating to impact and how to more effectively produce and disseminate impactful scenario products. A discussion on the impact of scenarios in general and reflections on such impact follows below.

2.4. The impact of scenarios

The previous section explored a number of South African scenarios in terms of their contents and impact. In terms of impact, Chantell Illbury and Clem Sunter refer to the "Wack" test, based on the ideas of Pierre Wack, a key scenarios planner in the 1970s and 1980s. According to this test, scenarios are not deemed important because of their prediction capability. What is important is their ability to influence the mindsets of decision-makers and to encourage them to act [39].

The issue of scenario impact is in many ways tied to a broader issue often referred to as the science-policy or science-end user interface. This issue essentially speaks to the challenge of getting knowledge that is produced by scientific or expert teams to be used in the public domain. This discourse recognises that there should be a close relationship between science or research products and their end-users, which could include government, policy-makers, businesses and communities. In reality, however, this relationship is not always an effective one, resulting in research often (or mostly) having minimal impact on policy and practice. The science-policy interface discourse explores why this happens in order to try to advise scientists and end-users about how to more effectively incorporate research into practice [40, 41, 42].

In terms of scenarios there tend to be two major opportunities for impact. The first is an impact on the participants who are part of the scenario development process. This is referred to as 'communication *for* scenarios' [43]. Similarly this opportunity for impact can be referred to as first order influence. First order influence refers to participants in the scenario development process undergoing personal changes in their thinking and behaviour. They also commit to the process, learn new skills, and build new networks and relationships. Because participants increasingly respect, understand and trust each other, they jointly commit to change [37].

The second is the impact of scenarios on broader society. This can be referred to as 'communication *of* scenarios' [43]. Here a wider group of stakeholders ideally need to be exposed to the scenarios once they are fully developed. As such, at this stage it is important to think about ways to foster appropriate dissemination and use of scenarios. This stage can also be referred to as second and third order influence. Second order influence is closely linked to first order influence. Participants who have been part of the scenario development process go back to their communities and networks and start sharing their new language, thoughts and insights with others. Third order influence is a process of social change, but can be difficult to monitor and study because of the many variable factors that influence every change process [37].

The following sub-sections reflect on the impact of scenario development on the participating team as well as the impact or influence of scenarios on broader society.

2.4.1. Impacts on participants in the scenario building process

Participants in a scenario development process actively engage and transform the process in the sense that they are asked to share their views, ideas, concerns and experiences in order to generate drivers to develop scenarios or stories from these drivers. It is important to recognise that this kind of individual impact is difficult to quantify and tends to be very subtle [35; 37]. Nonetheless, the kinds of impacts that individuals experience can include:

- Experiencing reframed mental models – By being forced, through the scenario development process, to articulate and share different perspectives and mental models, participants are made to think carefully about their perceptions and often re-think their views when faced with other participants' views and the need to move collectively towards a desired future [44].
- Gaining a broadened network of relationships – Scenario development processes bring together groups of people to have open and constructive conversations. This process fosters a shared understanding, trust and a sense of community [44].
- Regenerating energy, commitment, and action – By clarifying desired futures and building consensus about how different actions will navigate society towards certain scenarios, a sense of regenerated energy and commitment can be achieved. Also, with new commitment in place, new actions can be catalysed [44].
- Taking pride in participation – When interviewed, participants tend to be quite proud of their involvement in scenario development processes. This encourages them to use and share the learning from the scenario development process during other projects and/or engagements [35].
- Creating a common vocabulary, trust and mutual understanding – Through the process of developing scenarios these subtle processes tend to be fostered. This is important as it is through trust and understanding that people are able to work together towards a desired shared future [37].

2.4.2. Facilitating and forming a scenario team

Whilst it is clear that a subtle process of impact and transformation can occur in a scenario team, this does not happen automatically. There are a number of lessons that have been learned through scenario development processes over the years that need to be borne in mind.

Firstly, having a diverse team is important [45]. The team should come from different age, race and gender brackets as well as a wide range of ideological spectrums [35]. This diversity is important because the more diverse the team is, the more diverse the driver inputs will be and as such the richer and more accurate the scenario development process will be. Also, an inclusive rather than exclusive scenario development process lends legitimacy to the process [46].

Secondly, embracing transdisciplinarity in any scenario development process is important. This implies that in order for scenarios to have the impact they need, they should be produced by a team made up of multiple different actors from government, civil society, communities, and research institutions. This will help the team to take into account different types of knowledge that different actors have (such as technical, traditional, experiential, cultural, and political knowledge). In so doing the inherent complexity in future planning processes will be reflected [47].

Finally, working with a diverse team with different knowledge, experience and viewpoints is not always easy. Conflict can arise when participants with different viewpoints are made to work together. Also, meaningfully incorporating feedback from diverse sets of stakeholders tends to be a highly time consuming process. Given these and other challenges that can arise, the importance of having a skilled, sensitive and insightful facilitator cannot be underestimated. Such a facilitator needs to be able to manage strong individuals who dominate conversation with their own agendas, and needs to be able to encourage everyone to express their opinions during the scenario development process [35].

2.4.3. Impacts on broader society

The impacts of scenarios on broader society are harder to ascertain and measure than the impacts of scenario development on the scenario team itself. This is because there are no measurable criteria for quantifying the impact that scenario products have on society, be they in written or oral form. Also the outcome of scenario development processes can never be attributed to a single factor. Scenario development processes typically deal with broad developmental issues making the range of issues and actors that they try to affect diverse. Scenario development processes also happen within the context of a range of related social activities, such as developments in policy, civil society events and public debates. For example, in the case of developing the South African scenarios of the 1980s and 1990s, there were multiple social forums, political parties, and government groups working on transforming the country. These scenarios and their related processes were just one input amongst many others that were part of the broad transition process. Similarly, the Water Sector Institutional Landscape by 2025 scenarios exist alongside scenarios established by the National Water Resources Strategy, the various government departments that do strategic planning and forecasts in relation to water, and the host of grassroots organisations that work on managing water sustainably for the future. Any impact or change in the water sector must then be attributed to a whole range of interlocking factors rather than just one set of scenarios.

2.4.4. Facilitating the effective dissemination of scenario products to society

In order for scenarios to have influence in the broader public space a number of key lessons are important. Firstly, a broad and extensive communication process is a key requirement and should be planned and budgeted for from the beginning of the project [42]. It is important that such a process targets multiple different actors in society, and takes place at many levels of scale (local, provincial, national) [23, 41, 42] in order to engage society and attempt to create a better future [37]. Non-government actors are an important target audience because they are critical in terms of instigating social debate, bringing about grassroots changes and challenging authorities to improve their performance [42].

In government, actors need to be aware of scenario products and how they can make use of them [23]. With regard to the South African water sector in particular, there seems to be a need to enable officials from DWA to apply the outcome of scenarios thinking and processes in their strategic decision-making aimed at mapping out the future of the water sector. A possible way of enabling experts and government officials to think imaginatively and creatively about the future, given their considerable daily workload and challenges, would be to involve scenario experts as facilitators for strategic planning sessions. Such sessions should ideally take the form of one or two day workshops in order to remove government officials from their immediate working environment and enable them to apply their minds to thinking creatively and focusing exclusively on the planning task at hand [23]. When engaging with government departments, it is important to be sensitive to and aware of different issues inherent in the government hierarchy. Non-political, technical experts tend to have a good knowledge of technical issues, but it is also important to target more senior political actors as they tend to have more decision-making power and can therefore implement changes and ideas brought about through the scenario development process more effectively [42].

It has been argued that regardless of which actor is being focused on, there are three key points to bear in mind in terms of targeting actors with information. A clear plan of action needs to be laid out and followed up on. The information needs to be shared in a manner that is non-threatening, interactive and flexible. Scenarios can be disseminated by tapping into existing networks and events such as management meetings, seminars and the media [42].

In addition, the way that scenarios are packaged and communicated is important [42]. There is a whole host of ways that information can be packaged and disseminated. There can be face-to-face dissemination [23], where scenarios are verbally presented at workshops, conferences, public gatherings, business breakfasts, and corporate events. Style of presentation is crucial in this regard. The presentations need to be simple, clear and memorable. The presenter needs to be engaging and open to feedback from the audience [35]. Radio or television documentaries can also be utilised to disseminate scenario ideas and generate public debate [36, 40, 42].

Another option is to publish the scenarios in a written format. A range of media can be used. The scenarios can be published in books, illustrated pamphlets [23] and newspapers [48].

Cartoon artists can be brought on board to illustrate the scenarios. Magazines and web pages can also be targeted. Written documentation about scenarios has proved to be a successful model. For example, Sunter and Illbury's 'The Mind of the Fox: Scenario Planning in Action' [49] is popular reading material and widely distributed.

Finally, it is crucial that the scenario products are seen as legitimate from the start of the scenario development process. They need to have buy-in from influential people involved in the issue that the scenarios explore. This legitimacy is generated by ensuring that the facilitators of the process as well as the scenario team are respected. Although a range of actors must be included in the scenario development process, and must be targeted in the dissemination process, it remains important to include high level and well-connected people in the team as it is often these individuals who will provide the 'insider' links for scenarios to be heard and disseminated through channels of influence [42]. If these strategic individuals cannot be made part of the team, they need to be made aware of and kept informed about the scenario development process to secure their interest and support [35].

Dissemination is not without its challenges. It is challenging to disseminate in a way that suits and reaches a diverse audience with different languages, levels of education, varied professional backgrounds and cultures. Another challenge of ensuring the uptake of scenarios (and research in general) is that dissemination is often not part of the project planning process, and as a result funding often runs out before scenario uptake and use can be promoted [48]. Also, depending on how it is done, the dissemination of scenarios can be very expensive [35].

2.4.5. General reflections on impact

Over and above the specific processes linked to the impact of scenarios on the scenario team and broader society there are some general points that are important to bear in mind when planning for impact in relation to scenario products.

Firstly, when starting the scenario development process, it is important to be clear about the purpose of the process one is undertaking and designing it accordingly. What are the intended outcomes of the process? Who is the process meant to influence and what product(s) will be necessary for this to happen [35]? Essentially scenarios need to fill a strategic gap or opportunity in society [50].

Secondly, questions also need to be asked about the timing of the scenario development process. Is there likely to be sufficient recognition among the intended target audience(s) that the problem being addressed is important and that the process is therefore potentially beneficial? Is the political environment such that intended target audience(s) will be responsive to fresh, unorthodox thinking [35, 42]?

Thirdly, attention also needs to be paid to the legitimacy of those financing and promoting the process, and the credibility of the project team developing the scenarios in the eyes of both the sponsors and the target audience(s) [35, 40, 42].

3. Conclusions

In conclusion, it seems that since the initial High Road/Low Road scenarios were developed, scenario development has taken root in South Africa, with several follow up scenarios having been developed since [28]. This development suggests that South African decision-makers must deem scenario development to be of considerable importance and utility, as it is often government or government-related institutions that develop or commission new sets of scenarios. These subsequent scenarios seem to mirror their predecessors with their snappy titles and straightforward structure and certainly have the potential to inspire decision-makers with regard to their planning activities [28].

Based on the discussion and reflections above, scenario development should involve a focus on dissemination and impact from the onset of the scenario process. Impact can happen at the level of participants in the scenario development process as they are exposed to new ideas and start adopting a new way of thinking about current issues of importance. These ideas have the potential to slowly infiltrate the networks of these participants and to also influence their thinking. At the same time, it is important to have a strong dissemination process in order to reach as many people as possible beyond the project team. The High Road/Low Road presentations are an example of a highly effective dissemination process made possible by an engaging speaker and interesting topic that was clearly and simply brought across to a wide range of audiences. Another key method of dissemination is to raise awareness about where the scenarios can be found and to make it easy for people to access them. The open access route followed by the Dinokeng scenario team is a good example of a scenario document that is easily available on a website, accompanied by much useful background information. It is this dissemination phase that has been lacking in the Water Sector Institutional Landscape by 2025 scenarios, and a follow up process is needed to plan how more people could be made aware of these scenarios and their usefulness to decision-making and planning in the South African water sector.

It is also important to keep in mind that scenarios are likely to have a higher impact if they are developed with the intention of identifying or solving particular problems [51]. If there is an intended target audience with particular information needs from the beginning of the scenario process, the scenario team will be able to keep this in mind when developing the scenarios. This will also ensure more effective uptake of the scenarios as pre-defined end-users exist. In the water sector, for instance, it could be effective for decision-makers who are grappling with a particular issue to solicit scenario inputs to aid them in making decisions regarding that issue.

In terms of future research, three areas come to mind based on what has been discussed in this chapter:

Firstly, a large scale study (mostly comprising of interviews) is needed to understand in greater detail the impact of scenarios on scenario participants, society and government planning processes [28]. Much of what has been argued to date in terms of the impact of scenarios has been on the basis of inference and assumptions. It would be interesting, though admittedly also very difficult, to try to substantiate views around the impact of scenarios with empirical evidence.

http://dx.doi.org/10.5772/52577

Secondly, it would be important to study how a scenario team would know that the timing is right to come up with and disseminate a new set of scenarios. It is reasonably easy to see that scenarios would have been important for particular moments in history, for example the political transition in South Africa, but it is considerably more difficult to determine when there may be an ideal window of opportunity in future in which scenarios may make an impact. It may also be important to determine which factors other than and in support of ideal timing would be important for scenarios to achieve impact.

Thirdly, building on this chapter, it would be important to determine how best to ensure that scenarios can become more useful and practical to policy-makers and other end-users. How can scenario teams ensure that end-users know how best they may use scenarios in order to influence their future planning? The issue of providing navigation to and between different scenarios and future outcomes is important in this regard.

Clearly, scenario development is a useful process to help decision-makers cope with and plan amidst uncertainty. Particularly in the context of the South African water sector, it is important to recognise that uncertainty is deepening in many ways given the impending presence of multiple stressors such as climate change, basin closure, growing populations, migration movements and a growing economy. These stressors, along with the institutional fluctuations and changes within the water sector itself, make it increasingly important for decision-makers to work with scenarios to help them to plan sensibly and creatively despite uncertainty. However, in order for scenarios to be useful it is important to plan for and carefully think about how to maximise their impact.

Acknowledgements

The authors would like to acknowledge the CSIR's librarian, Engela van Heerden, for her excellent work in sourcing a large amount of relevant literature that contributed to this chapter. They would also like to acknowledge Wilma Strydom for her valuable review comments on this chapter. Finally, the authors would like to acknowledge the Water Research Commission (WRC), the organisation that funded the development of the Water Sector Institutional Landscape by 2025 scenarios. The learning that the project team gained from this project was a key input into this chapter.

Author details

Nikki Funke, Marius Claassen and Shanna Nienaber

*Address all correspondence to: nfunke@csir.co.za

Natural Resources and Environment Unit, Council for Scientific and Industrial Research, Pretoria, South Africa

References

[1] Bogardi JJ, Dudgeon D, Lawford R, Flinkerbusch E, Meyn A, Pahl-Wostl C, Vielhauer K and Vörösmarty C. Water Security for a Planet Under Pressure: Interconnected Challenges of a Changing World Call for Sustainable Solutions. Current Opinion in Environmental Sustainability 2012; 4 35–43.

[2] Slaughter RA.Welcome to the Anthropocene. Futures 2012; 44 119–126.

[3] WWAP (World Water Assessment Programme). The United Nations World Water Development Report 4: Managing Water under Uncertainty and Risk. Paris: UNESCO; 2012.

[4] UN (United Nations). Declaration of the United Nations Conference on the Human Environment. Stockholm, 5-16 June 1972.

[5] Brundtland GH. Address by Mrs Gro Harlem Brundtland, Chairman at the Closing Ceremony of the Eighth and Final Meeting of the World Commission on Environment and Development 27 February 1987. Tokyo, Japan.

[6] WMO (World Meteorological Organisation). International Conference on Water and the Environment: Development Issues for the 21st Century. Geneva: ICWE Secretariat; 1992.

[7] UN (United Nations). Agenda 21. United Nations Conference on Environment and Development Rio de Janeiro, Brazil, 3 to 14 June 1992.

[8] Spangenberg JH, Pfahl S, Deller K. Towards Indicators for Institutional Sustainability: Lessons From an Analysis of Agenda 21. Ecological Indicators 2002; 2 61–77.

[9] UN (United Nations). World Development Report, 1980. Washington, D.C.: The World Bank; 1980.

[10] UN (United Nations). World Development Report, 1985. New York: Oxford University Press; 1985.

[11] UN (United Nations). World Development Report, 1990: Poverty. New York: Oxford University Press; 1985.

[12] UN (United Nations). World Development Report, 1995: Workers in an Integrating World. New York: Oxford University Press; 1995.

[13] World Bank. Indicators. http://data.worldbank.org/indicator. (accessed 30 July 2012).

[14] Bates C, Kundzewicz ZW, Wu S and Palutikof JP., editors. Climate Change and Water. Technical Paper of the Intergovernmental Panel on Climate Change. Geneva: IPCC Secretariat; 2008.

[15] Davis CL. Climate Risk and Vulnerability: A Handbook for Southern Africa. Pretoria: Council for Scientific and Industrial Research; 2011.

[16] Tewari DD. A Detailed Analysis of the Evolution of Water Rights in South Africa: An Account of Three and a Half Centuries from 1652 AD to the Present. Water SA 2009; 35(5) 693-710.

[17] Gleick PH. The World's Water, 1998-1999. Washington DC: Island Press; 1998.

[18] DWAF (Department of Water Affairs and Forestry). National Water Act of South Africa. Act No. 36 of 1998. Pretoria: Department of Water Affairs and Forestry; 1998.

[19] DWAF (Department of Water Affairs and Forestry). National Water Resources Strategy – First Edition. Pretoria: Department of Water Affairs and Forestry; 2004.

[20] DWA (Department of Water Affairs). Annual Report of the Department of Water Affairs Vote 37: 1 April 2010 To 31 March 2011. http://www.pmg.org.za/minutes/30. (accessed 6 August 2012).

[21] Emerson JW, Hsu A, Levy MA, de Sherbinin A, Mara V, Esty DC, Jaiteh M. Environmental Performance Index and Pilot Trend Environmental Performance Index. New Haven: Yale Center for Environmental Law and Policy; 2012.

[22] Green GC, Day JA, Mitchell SA, Palmer C, Laker MC, Buckley CA. Water Research Commission 40-Year Celebration Conference - Syntheses of Themed Sessions. Water SA 2011; 37(5) 609-618.

[23] Claassen M, Funke N, Nienaber S. 2011. The Water Sector Institutional Landscape by 2025 Technical Report. WRC Report No. 1841/1/11. Pretoria: Water Research Commission; 2011.

[24] Mietzner D, Reger G. EU-US Seminar: New Technology Foresight, Forecasting and Assessment Methods. Seville; 13-14 May 2004.

[25] Landeta J. Current Validity of the Delphi Method in Social Sciences. Technological Forecasting and Social Change 2006; 73 467-482.

[26] Daum J. How Scenario Planning Can Significantly Reduce Strategic Risks and Boost Value in the Innovation Chain. The New Economy Analyst Report 8 September 2001.

[27] Sunter C. The World and South Africa in the 1990s. Cape Town: Human Rousseau Tafelberg; 1987.

[28] Galer G. Scenarios of Change in South Africa. The Round Table 2004; 93(375) 369-383.

[29] DACST (Department of Arts, Culture, Science and Technology). South African National Research and Technology Foresight Project. Pretoria: DACST; 1999.

[30] Dinokeng team. 3 Futures for South Africa. www.dinokengscenarios.co.za (accessed 10 August 2012).

[31] (WBCSD) World Business Council for Sustainable Development. Business in the World of Water: WBCSD Water Scenarios to 2025. www.wbcsd.org/web/H2OScenarios.htm (accessed 30 July 2012).

[32] GRA (Global Research Alliance) Science and Technology-Based Water Scenarios for Sub-Saharan Africa. www.research-alliance.net. (accessed 6 August 2012).

[33] Peterson GD, Cumming GS, Carpenter SR. Scenario Planning: A Tool for Conservation in an Uncertain World. Conservation Biology 2003; 17(2) 358-366.

[34] Finlev T. Future Peace: Breaking Cycles of Violence through Futures Thinking. Journal of Futures Studies 2012; 16(3) 47-62.

[35] Segal N. Breaking the Mould: The Role of Scenarios in Shaping South Africa's Future. Stellenbosch: SUN Press; 2007.

[36] Kahane A. The Mont Fleur Scenarios: What Will South Africe Be Like in the Year 2002? Deeper News 1992, 7(1) 1-22.

[37] Maddison S, Cronin D, Williams S, Coggan R. Democratic Dialogue: Finding the Right Model for Australia. Indigenous Policy and Research Unit, Discussion Paper No.1. Sydney: University of New South Wales; 2009.

[38] Claassen M, Funke N, Nienaber S. The Water Sector Institutional Landscape by 2025. WRC Report No. TT 514/11. Pretoria: Water Research Commission; 2011.

[39] Karuri-Sebina G, Rosenzweig L. A Case Study on Localising Foresight in South Africa: Using Foresight in the Context of Local Government Participatory Planning. Foresight 2007, 14(1) 26-40.

[40] Strydom WF, Funke N, Nienaber S, Nortje K, Steyn M. Evidence-based Policy-making: A Review. South African Journal of Science 2010; 106(5/6) 249 – 334.

[41] Funke N, Nienaber S, Henwood R. Scientists as Lobbyists? How Science Can Make its Voice Heard in the South Africa Policy-Making Arena. International Journal of Public Affairs 2011; 10(102) 421 – 443.

[42] Funke N, Nienaber S. Promoting Uptake and Use of Conservation Science in South Africa by Government. Water SA 2012; 38(1) 105-114.

[43] Lebel L, Thongbai P, Kok K, Agard JBR, Bennett EM, Biggs R, Ferreira M, Filer C, Gokhale Y, Mala W, Rumsey C, Velarde SJ, Zurek M, Blanco H, Lynam T, Tianxiang Y. Sub-global Scenarios. In: Capistrano D, Lee M, Raudsepp-Hearne C, Samper C. (eds.) Ecosystems and Human Wellbeing: Multi-scale Assessments. Findings of the Subglobal Assessments Working Group of the Millennium Ecosystem Assessment. Washington D.C.: Island Press; 2005. p229-259.

[44] Ringland G. Scenario Planning, 2nd edition. England: John Wiley & Sons; 2006.

[45] Audouin M, Preiser R, Nienaber S, Downsborough L, Lanz J, Mavengahama S. Exploring the Logic of Complexity for Researching Social-Ecological Systems. Ecology and Society In Press.

[46] Ogilvy JA. Creating Better Futures: Scenario Planning as a Tool for a Better Tomorrow. New York: Oxford University Press; 2002.

http://dx.doi.org/10.5772/52577

[47] Jacobs I, Nienaber S. Water Without Borders: Transboundary Water Governance and the Role of the 'Transdisciplinary Individual' in Southern Africa. Water SA (WRC 40 Year Celebration Special Edition 2011; 37(5) 665 – 678.

[48] Communication with Clem Sunter. 26 July 2012. Pretoria.

[49] Illbury C, Sunter C. The Mind of a Fox. Cape Town: Human & Rousseau/Tafelberg; 2001.

[50] Johnson G, Scholes K, Whittington R. Exploring Corporate Strategy: Text and Cases. New Jersey: Prentice Hall, Financial Times Press; 2008.

[51] Bohenksy E, Reyers B, Van Jaarsveld AS. Future Ecosystem Services in a Southern African River Basin: a Scenario Planning Approach to Uncertainty. Conservation Biology 2006; 20(4) 1051-1061.

Chapter 2

Recent Developments in Monitoring of Microbiological Indicators of Water Quality Across a Range of Water Types

Sandra Mesquita and Rachel T. Noble

Additional information is available at the end of the chapter

http://dx.doi.org/10.5772/45847

1. Introduction

As human pressure increases, so does exploitation of natural water sources for production of drinking water, public use and appreciation of recreational waters. Subsequently, the demand for faster and reliable monitoring methods and approaches has also been intensified. Ensuring that water is safe, whether for bathing or consumption is a critical complex process that requires the participation of multiple stakeholders [1, 2] and also distinct degrees of knowledge with respect to water management.

For quantifying microbial contaminants that are dangerous to public health, fecal indicator bacteria (FIB) are used worldwide as indicators of the potential presence of dangerous pathogens that can be found in water used for bathing, drinking and harvesting of seafood. The "indicator" system relies on the assumption that FIB are present concomitant to the presence of human bacterial, viral and parasitic pathogens of concern. Furthermore, the indicator system simplifies management, because no single pathogen can be singularly attributed to the majority of waterborne disease. If pathogens were measured for management of water quality, the list of candidates to be quantified would be many. The FIB system relies on thresholds that have been established using epidemiological studies and risk assessment frameworks (see Annapolis Protocol and WHO bathing water guidelines).While the standards for drinking water, recreational water and shellfish harvesting waters and meats are based upon similar approaches [3], different types of waters can have different acceptable thresholds for the same FIB because of differences in either acceptable risk, the relative use of the water (i.e. for consumption versus recreation), and a range of other factors. Therefore, acceptable thresholds for FIB in drinking water, recreational water and shellfish harvesting waters and meats vary by continent, region, state and locality.

By definition an indicator (in this case, FIB) should be, 1) able to be enumerated using standardized tests that are relatively cost-effective and user-friendly, 2) present in high concentrations in human fecal contamination, matching the high concentrations of other respective pathogens, and 3) well studied in relation to human health, with epidemiological studies to link FIB concentrations and adverse health outcomes such as gastrointestinal and respiratory human diseases [4, 5, 6]. These assumptions can be tenuous at times for all types of waters and for all of the FIB groups currently utilized for water quality management. However, it is generally accepted that regulation using FIB to protect public health is successful. For example, total coliforms do not exactly fit these requirements since they are not exclusively fecal in origin; however they are often used in conjunction with other FIB enumeration approaches to protect public health in drinking water [7]. Even though some of the assumptions listed above are violated periodically for the use of FIB to protect public health in the regulation of drinking water, an alternative approach is often unattainable. Human pathogens are heterogeneous, can be highly seasonal, and can be highly infectious at low doses. The range of enteric viruses alone that could be monitored in drinking water is vast, and these pathogens are difficult to detect [8, 9, 11]. A discussion of the pathogens of concern is beyond the scope of this chapter, and the reader is referred to a range of quality publications on the topic (e.g. 11, 12).The "indicator" system can be used successfully, particularly if new research and methodologies in the water quality arena are incorporated into management. As a community, we can attain improved means of management of public health in water, if attention is paid to utilization and incorporation of new monitoring tools, new models and molecular tools. Current advances in science include rapid quantitative PCR assays (QPCR) developed to track distinct microorganisms in multiple water sources [12, 13, 14, 15, 16, 7].

In this chapter, we highlight recent advancements in the management of water, and in monitoring of microbial water quality in drinking water production and recreational waters. First, we highlight the current regulatory landscape for microbiological indicators. Second, we will describe the application and evolution of molecular tools for rapid quantification of FIB (E. coli and *Enterococcus* sp.), in a range of water types. We will include some of the limitations and advantages of different types of culture based methods as compared to the recently developed rapid QPCR-based methods. We will present a small case study conducted along high use beaches along the Pacific Ocean of California, USA. Third, we highlight recent developments in the field of Harmful Algal Blooms (HABs) and concerns over cyanobacterial toxins in surface water sources for drinking water. Following the presentation of information presented on HABs, and comparing *E. coli* and *Enterococcus*sp. concentrations using culture based and molecular (qPCR) rapid methods, the authors will consider the application of newly developed source tracking markers for quantification of fecal contamination fecal contamination. These microbial source tracking (MST) tools are currently being used and expanded for the management of a range of water types. The chapter is intended to highlight recent advancements, while promoting the consideration of other successful ap-

http://dx.doi.org/10.5772/45847

proaches for managing water resources well into the 21st century with an eye toward improved protection of public health.

2. Regulatory landscape across major water types: Highlighting the use of the fecal indicator bacteria as part of the "indicator" system

Traditionally, the application of FIB for water quality monitoring included enumeration of members of the total coliform group. This group was originally described on the basis of lactose fermentation detection instead of upon the principles of systematic bacteriology [3; 17-26]. Total coliforms are still used widely for management of drinking water, but their sole use in recreational water quality monitoring has decreased in favor of enumeration of FIB that are specific to fecal contamination from warm-blooded animals (i.e. *E. coli* and *Enterococcus* sp.). Total coliforms are defined by their respective characteristics, i.e. they are Gram negative facultative anaerobes that do not form spores are rod-shaped bacteria, with lactose fermentation occurring with acid production (24-48h, 36°C), and are indole-negative. Coliforms belong to the family Enterobacteriaceae which includes *Escherichia, Enterobacter, Klebsiella* and *Citrobacter, Kluyvera, Leclercia,* and some members of the genus *Serratia* [27].

In all of the culture-dependent methods, cultivation conditions are choose in order to improve the growth of the target microorganism while simultaneously inhibiting the growth of other microorganisms. Balance amongst sensitivity and selectivity is the reason for different methods for sample processing (drinking water *vs* highly contaminated waters) [28] and selection of different approaches for quantification using traditional culture based methods. The three most widely used culture based methods for quantification of FIB in any water type are Defined Substrate Technology (DST, e.g. IDEXX Colilert-18®), multiple tube fermentation (MTF), and membrane filtration (MF). The first edition of "Standard Methods for the Examination of Water and Wastewater" was released in 1905 and *E. coli* was selected as the most suitable indicator organism for raw drinking water [29]. Therefore multiple detection methods for *E. coli* were available early in the century. As scientific knowledge evolved a broader group of microorganisms (still including total coliforms, fecal coliforms, and *E. coli*) have been selected as surrogate measures in a wide range of water sources and treated water [30]. Two key factors have led to the trend of using *E. coli* as the 'preferred' indicator for the detection of fecal contamination, not only in drinking water but in other matrices as well: first, the finding that some 'fecal coliforms' were non-fecal in origin, and second, the development of improved testing methods for *E. coli* [1-19]. Membrane filtration is commonly selected for FIB quantification along the drinking water treatment process since it is flexible to the amount of sample filtered and since the method permits specific quantification. That is, both MTF and DST-based methods are either reported in a presence/absence format or in a "most probable number" format. Colilert® and Colilert- 18®, however, tend to be used at the "end of pipe" and are user friendly methods, such that their popularity has increased in the past decade.

An example of current regulations for protection of public health in Europe, USA and other countries is summarized in Table 1 [20]. Note that many currently used methods both total coliform and *E. coli* enumeration together, as IDEXX DST kits such as Colilert-18® permit the simultaneous quantification of both groups. Furthermore, while guidelines have been issued in different countries, as highlighted in Table 1, the interpretation and implementation of monitoring typically falls in the hands of member states and provinces; this creates a wide range of monitoring approaches that are currently employed for management of drinking, recreational, and shellfish harvesting water quality. It is for this reason that only a table for drinking water is presented, as drinking water regulations are the most stringent and consistent across nations.

Parameter	Canada	United States	United Kingdom	EU directive	WHO*	Australia
Total Coliforms	0/100mL in 90%	0/100mL in 95%	0/100mL	0/100mL	*	*
Thermotolerant coliforms (fecal coliforms) or *E. coli*	*	0/100mL	*	*	0/100mL	*
E. coli	0/100mL	0/100mL	0/100mL	0/100mL	0/100mL	*
Enterococci	*	*	0/100mL	0/100mL	*	*
Cryptosporidium parvum	*	99% removal or inactivation	<1oocyte/ 10L	*	*	*
Cryptosporidium perfringens (including spores)	*	*	0/100mL	a	*	*
Pseudomonas aeruginosa	*	*	*	0/250mL	*	*
Colony count 22°C	*	*	No abnormal change	No abnormal change	*	*
Colony count 37°C	<500CFU/mL [b]	<500CFU/mL [b]	No abnormal change	20/mL [c]	*	*
Microcystins _LR	1,5µg/L	*	*		1,0 µg/L (provisional)	Lifetime exposure 1,3µg/L

* spaces left in blank indicate parameters not specified; [a] Necessary only if the water originates from or is influenced by surface water;

[b] HPC(35°C for 48h) or <200 background coliforms on a total coliform membrane filter; [c] Necessary only in the case of water offered for sale in bottles or container

Table 1. International drinking water standards and guidelines from the World Health Organization (WHO) Modified from [31, 32 and 33]

The World Health Organization (WHO) Guidelines for Drinking-water Quality [32] represent an overall international scientific consensus, based on a wide range of participants, of the health risks presented by microbes and chemicals in drinking water. The existence of international guidelines for cyanobacteria microcystins (1 μg/L microcystins –LR) in drinking water has been established by WHO (1998). Most countries rely upon this value and developed specific systems adapted for their own reality. For instance, in Brazil federal legislation requires more intensive monitoring programs including toxin analyses or toxicity testing if cyanobacteria exceed 10,000 cells/ml or 1 mm^3 biovolume in a given water sample. This includes a mandatory standard of 1 μg/L applied for microcystins (variants not specified), and recommendations limiting to saxitoxin 3μg/L.

Evaluation of water quality for drinking water production requires sampling collection at specific temporal intervals to ensure safety for the public. Usually it is recommended to sample across the process of drinking water treatment, permitting an analysis of bacteriological water quality throughout processing and distribution [6,7]. This permits rapid identification of a process or component in the distribution system that is failing, and rather than troubleshooting across the entire process, subsequent attention can be focused on specific components. This approach simultaneously prevents local contamination from developing undetected at cross-connections or breaks in the distribution lines or due to a drop in positive pressure [1, 34, 35].

The definition of a recreational water body represents a wide range of environments that include the ocean, hot springs, lakes, reservoirs, streams and rivers [34]. Similarly, the diversity of potential human pathogenic microorganisms includes all viruses, protozoans, and bacteria that could potentially be present in natural fresh and marine recreational waters particularly those contaminated by wastewater [9, 36]. In recreational water quality monitoring, monitoring approaches are often guided by use. Beaches that are used intensely by the public (such as those in Santa Monica Bay, California, USA) are monitored daily throughout the summer bathing season. In some states or areas with lower beach usage, during specific "off seasons", or areas that are used only for secondary contact recreation such as kayaking and sailing, monitoring for FIB might be conducted only once per month. Although a broad discussion is beyond the scope of this document, the use of FIB for monitoring shellfish harvesting waters also relies heavily on the use of traditional culture-based methods such as MF and MTF, but management differs by country. In some EU states, management of shellfish is by quantities of *E. coli* in shellfish meats (e.g. oysters), whereas, quantification in the United States is based upon fecal coliforms in the water column surrounding the shellfish beds.

Advantages connected to the application of culture-based methods include low equipment requirements, user-friendliness, and low cost, which makes it the most used approach for quantifying FIB in any water type. On the other hand, the enrichment stage takes time (typically from 18-24 hours) causing the notification of the public to be delayed during a contamination event [37]. This delay also causes vital revenue to be lost because contaminated waters are often still restricted or "closed" for use (in the case of beaches) after the contamination event has already passed. Incubation methods also suffer from the fact that they can promote the growth of false positive organisms [38]. Rapid molecular methods have recently been developed for application to protection of public health in recreational waters [39]. These methods require 3 hours from sample processing to results, and they are qPCR based. They target specific organisms (*Enterococcus* sp. or *E. coli*). The use of such methods permits rapid public notification (e.g. [40]) resulting in improved protection of public health. One other advantage of such methods is the high specificity [28X]. Other rapid methods using qPCR have already been developed for quantification of microbial markers that are specific to sources of fecal contamination [44, 45].

3. Newly developed rapid molecular methods for recreational water quality monitoring

Millions of dollars are expended each year to measure FIB to and assess whether swimming at recreational beaches posed a risk to public health. The monitoring programs are typically conducted using culture-based methods that require an 18 to 96 hour incubation period. The lengthy incubation step results in situations where beaches are managed inaccurately, by keeping beaches open that are contaminated and closing beaches that are potentially clean again [40]. Newer, more rapid methods, such as qPCR have been developed by EPA and others for use in recreational water quality monitoring [39, 41]. Several of these new rapid PCR-based methods for quantification of *Enterococcus* sp. have been found to be significantly related to human health outcomes in epidemiology studies [42, 43]. In some cases, the association between rapid methods and adverse human health outcomes has been stronger than that observed between traditional culture-based methods and adverse human health outcomes [43].

Quantitative PCR is a novel primer-based molecular technique that combines the specificity of traditional PCR with the quantitative measurement of fluorescence for quantification of presence of specific types of nucleic acid in environmental samples. Many of the recently developed methods are actually quite similar to one another; differing only in the type of qPCR fluorophore "chemistry" used [39,41]. Even though epidemiological studies to assess relationships between rapid FIB quantification methods and human health outcomes have been successful, the implementation or widespread use of the rapid qPCR-based methods has not been without hurdles. Noble et al., 2010 [41] successfully demonstrated the use of rapid molecular methods by water quality personnel that had no previous experience with molecular techniques. Griffith and Weisberg, 2011 [40] built upon this success, by training three water quality laboratories to conduct the rapid methods, and implementing the use of the rapid methods for active beach management decisions (posting and closing of beaches). They also demonstrated the use of real time notification technology at the beaches to rapidly convey real-time water quality results through the assistance of a non-profit organization called "MiOcean". Griffith and Weisberg (2011) reported excellent management agreement (>96%, i.e. the agreement in whether a beach would be posted or remain open) among the rapid QPCR based enumeration of *Enterococcus* sp. [41] and culture based methods such as EPA Method 1600 (membrane filtration based quantification for *Enterococcus* sp.). Lavander and co-workers [46] have shown similar successes with quantification of *E. coli* at freshwater

Great Lakes beaches, reporting management agreement of 98% between the rapid qPCR based quantification of *E. coli* and traditional culture based methods such as Colilert-18.

Even though success has been demonstrated in specific environments, it is clear that rapid qPCR-based methods for quantification of FIB are not appropriate for all places at all times. Three major issues to be resolved are ways to standardize assays, the approach used for quantification of the target organism, and inhibition of the qPCR.

Currently there are two modes of quantification being employed in the literature, both with advantages and disadvantages. One relies on quantifying the target cell, e.g. *Enterococcus* sp., where results are reported as cell equivalents/100 ml, where inhibition is assessed using a specimen processing control, but the final reported target concentrations are not modified quantitatively using inhibition data (see [40] for an example). The second approach reports target cell concentrations as calibrated cell equivalents/100 ml and relies on the use of cycle threshold ratios for both the target cell and the specimen processing control for final calculated quantities of the target cell concentration. The cell equivalent approach can, in certain circumstances, provide data that is more related to historical data reported for culture based methods [40,41, 46], but cannot be used at beaches were partial inhibition of the qPCR is common. On the other hand, the calibrated cell equivalent quantification approach can be more widely applied at a range of beaches, because inhibition is accounted for. However, the approach is imperfect because the specimen processing controls that are currently used are DNA-based, and they do not adequately predict inhibition of the amplification of FIB cells in real world samples [46].Dilution is another approach that can be used to effectively reduce inhibition, but this increases the limit of detection.

Figure 1 is a snapshot of results from a preliminary assessment of the performance of rapid qPCR-based methods as compared to membrane filtration (culture-based) methods for the quantification of Enterococcus sp. in water samples collected from an array of beach locations in Santa Monica Bay, California, USA. The rapid methods used here provide results within a few hours of sample collection. Quantification was conducted using the cell equivalent approach.

Inhibition of the qPCR has been discussed at length in other publications but is relevant to the implementation of rapid qPCR based methods for water quality assessment because of its unpredictable nature. Inhibition of qPCR can occur when high molecular weight compounds in the surface water (e.g., humic acids and other complex carbohydrates) combine with metal ions to sequester nucleic acids from polymerases and prevent amplification [47, 48, 49, 50]. Inhibition can sometimes be alleviated with the use of a commercial DNA extraction kit for sample purification. This requires extra time that is added to the total analysis time, and increases variability in the final results due to analyst error and variable DNA binding characteristics. Others have tried to deal with inhibition in analysis of surface and other water samples with the addition of adjuvants (e.g. bovine serum albumin, [49]. In recreational waters, many use salmon testes DNA to assess inhibition (e.g. 39). Cao et al. 2012 [10] assessed the use of dilution, salmon testes DNA, and internal controls for the assessment of inhibition in a range of water sample types, and no one approach clearly emerged as superior over the others. In fact, the internal controls and salmon testes DNA did not agree in their assessment of inhibition of the target qPCR [52]. It may be that implementation of the rapid methods, at least at first, follows the approach developed in southern California by [40]. They utilized the cell equivalent quantification approach, and for beaches that demonstrated inhibition of the QPCR, traditional culture based methods were utilized to quantify FIB concentrations for beach management purposes.

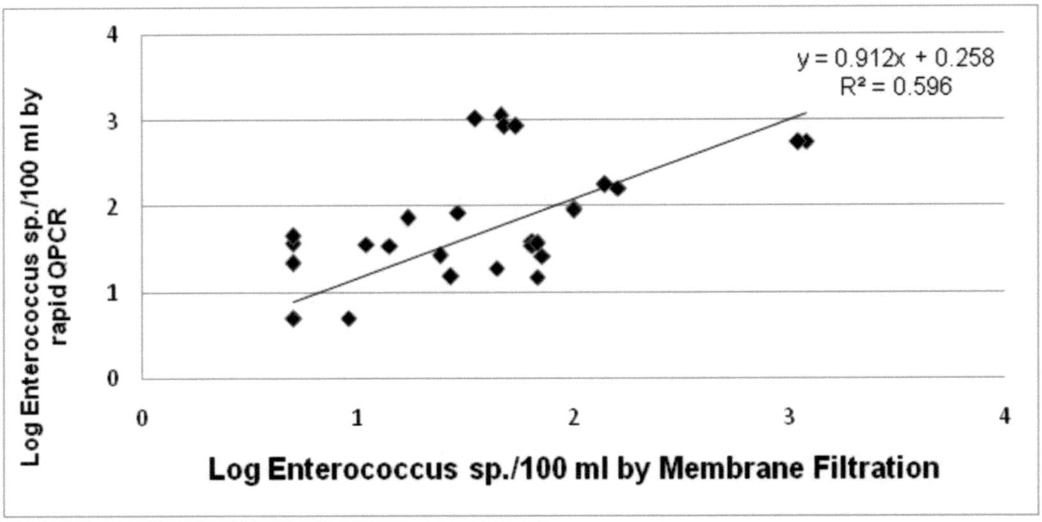

Figure 1. Example of rapid qPCR-based methods compared to membrane filtration (culture based) methods for the quantification of Enterococcus sp. in water samples collected from an array of beach locations in Santa Monica Bay, California, USA.

In the past years another new, rapid and reliable methodological approach has been tested to monitor microbial contamination, particularly in recreational marine and freshwater sources [53, 54]: the immunomagnetic separation / adenosine triphosphate (IMS / ATP) method. IMS-ATP was first developed by [55] and consists of the use of immune magnetic beads that selectively capture targeted bacteria using cell specific antibodies. A magnet is then used to separate the target bacteria from the rest of the sample with concentration determined by quantification of the ATP concentration in the sample using a bioluminescent assay [53]. While IMS/ATP has not been the subject of potential recommendation by the USEPA in 2012, there are advantages to its use. These advantages include the detection of only viable bacteria, therefore making it more directly related to quantification based upon other currently used culture based methods. Rapid quantification using IMS/ATP might be beneficial in areas receiving treated, disinfected wastewater for example, where the concentration of FIB-specific DNA is likely high, but the concentration of viable FIB cells would be lower. The capital costs of IMS/ATP are lower than those of the molecular methods. Another positive aspect of this technique is that it requires less than 4 hours. The major criticism of IMS/ATP concerns the availability of appropriate antibodies for quantification, particularly for the *Enterococcus* genus, which cannot be adequately captured using a single set of antibodies. However, improvements are being made in the development of cocktails of antibodies combined with the development of new bead based capture systems [e.g. 54].

In the coming years, there is the potential to dramatically improve the sensitivity, accuracy, precision, and speed of the QPCR assays used for water quality applications, while dramatically reducing the cost associated. One key step forward will be in the area of standardization, as standardized kits and cell standards and controls become available in ready-made kits, the analyst error and variability associated with the assays will decrease dramatically. The idea is to develop an elegant and powerful, yet simple, tool that can be applied to a wide variety of water sample types, to simultaneously determine the FIB concentration and enumerate bacterial tracers or pathogen types, necessary for preliminary source identification and/or protection of public health. Many key groups around the globe are working collaboratively to ensure this success.

4. Harmful Algae Blooms (HABs) in drinking water treatment plants and recreational waters

In general, harmful algal blooms (HABs) are a cause of great concerns amongst stakeholders involved in water quality regulation and management. Since this chapter is specifically devoted to the problems caused by microbial contamination, we are specifically focusing our treatment of the subject on the closely related cyanobacteria, and their ability to form toxic blooms (referred to herein as CYANOHABs). There is an array of reports on the increased incidence of toxic blooms caused by Cyanobacteria. Their cyanotoxins are a serious concern for raw water sources that are exploited for drinking water production and recreational purposes [33]. Most countries around the world rely upon the exploration of surface freshwater sources for drinking water production. Efficient surface water treatment systems usually combine a series of treatment stages. Commonly employed water supply treatment includes multiple steps, generally including in order: pre-chlorination, flocculation and coagulation (addition of salts of iron or aluminium), sedimentation and post-chlorination [R]. One specific problem associated with toxic blooms of cyanobacteria is that specific stages of drinking water treatment are needed to remove the toxins. Not all of the drinking water treatment plants have these stages available or possess a flexible system [56]. The goal is to guarantee the production of drinking water according to national water quality guidelines, including microbial and chemical parameters that could potentially promote human disease and/or fatalities. In Table 2 some of the international guidelines for the presence of cyanobacteria toxins is presented. Cyanotoxins can be classified from to a toxicological perspective: hepatotoxins, cytotoxins, neurotoxins, irritant compounds as well as dermatotoxins [33]. Common methods of detecting and quantifying cyanotoxins include Enzyme-linked immunosorbent assay (ELISA), and chromatographic analysis [33, 57].

The main concern for public health associated with cyanobacterial blooms is exposure to cyanotoxins which can cause central nervous system or liver damage. There are multiple modes of exposure to cyanotoxins but most common are: ingestion of contaminated water, accidental ingestion by swimming on contaminated waters, and consumption of vegetables irrigated by contaminated waters or even inhalation (Table 3) [33,57]. A CYNOHAB's cause adverse effects to drinking water production plants. For instance there are physical impacts and the release of hazard by-products from the treatment stages such as chemical treatment. Other hazards come from oxygen depletion or pipe corrosion due to microbial activity within the pipes coming from bloom collapse. It is fundamental to have knowledge of source waters such as lakes that are prone to the development of toxic and non-toxic cyanobacterial blooms so that the overall population can be informed about public health risks. Several statistical approaches have been used to create models that could explain phytoplankton feed-back to nutrient enrichment and oxygen reduction, and these models may be adapted to other locations [56,58].

Whilst WHO's guidelines are based on total of enumerated cells (using microscopy techniques) and/or determination of chla (chlorophyll a) concentrations related to the amount of cells, there is a clear desire to develop faster quantification of cyanotoxins in environmental samples, especially large population areas. It is important to have rapid, reliable and accurate analysis i.e. high throughput molecular detection methods. In order to develop molecular assays, pure cultures from toxin producers need to be utilized to focus on genes specifically involved in the synthesis of a given toxin [AS].In Table 2, the guidelines from WHO for cyanobacterial levels and expected health effects due to exposure of contaminated recreational waters are shown.

Guidance level or situation	How guidance level derived	Health risks
20000 cyanobacterial cells/mL 10μg chlorophyll-a/L Low impacts in human health	From human bathing epidemiological study	Skin irritations and gastrointestinal illness
100000 cyanobacterial cells/mL or 50μg chlorophyll-a/L moderate chances of affecting human health	From provisional drinking water guideline value and data concerning other cyanotoxins	Long-term illness and/or Short-term problems such as skin irritations and gastrointestinal illness
High probability of adverse effects Scum formation in areas where whole-body contact and/or risk of ingestion/aspiration occur	Inference from oral animal lethal poisonings	Potential for acute poisoning
	Actual human illness case histories	Potential for long-term illness with Adverse health problems such as skin irritations and gastrointestinal illness

Table 2. WHO guidance values for recreational waters with respect to cyanobacteria contaminates waters.(modified from [33]).

Recently published molecular approaches for detection and quantification of toxigenic and non-toxigenic cyanobacteria are based on the application of multiplex approaches for targeting the genus (16S rDNA), and subsequently using other nested assays for sequences found in the non-ribosomal polypeptide synthetize complex (55KB in size) responsible for toxin production [56,57, 58]. Al-Tabrineh et al. (2012) recently developed and validated a quadruple target qPCR assay specifically designed for the detection and quantification of genes involved in the production of microcystins, nodularins, cylindrospermopsins and saxitoxins. There is a multiplicity of qPCR protocols published recently; however they typically focus on the quantification and detection of *Microcystis* spp. as a target due to the fact that it represents the most recurrent blooming cyanobacterial group in multiple water sources around the globe [59].

CYANOHAB related problems begin at the raw water source. Therefore, international guidelines such as the Water Frame Directive (Water Frame Directive, WFD; Directive 2000/60/CE of 23 October 2000) require a good ecological state in freshwater or highly modified freshwater sources (example water dams). Therefore, if a water body doesn't comply with these requirements it should be considered contaminated treated accordingly adding necessary steps to guarantee good finished water quality [60]. Thus, regular monitoring of freshwater sources for both recreational and drinking water purposes are necessary. Current modes of quantification starts with chla, but can extend to molecular quantification of specific phytoplankton groups, quantification of toxins (typically ELISA based), or pigment based quantification of phytoplankton groups [61].

Cyanotoxins	**Mechanism of toxicity**	**Taxa toxin producer**	**LD_{50}(mouse) pure toxin**
Protein-phosphatase blockers			
General microcystins	Blocks protein phosphatases by covalent binding promoting liver haemorrhage damage could occur	Microcystis, Planktothrix, Anabaena, Oscillatoria, Nostoc, Anabaenopsis, Hapalosiphon, Nodularin	45->1000µg/Kg
Microcystin-LR			60(25-125)µg/Kg
Microcystin-YR			70 µg/Kg
Microcystin-RR			300-600 µg/Kg
Nodularin			30-50 µg/Kg
Neurotoxins			
Anatoxin-a	Blocks postsynaptic depolarization	*Aphanizomenon, Oscillatoria, Anabaena, Cylindrospermum*	250 µg/Kg
Saxitoxins	Blocks sodium channels	*Aphanizomenon, Anabaena, Lyngbya, Cylindrospermopsis raciborskii*	10-30 µg/Kg
Cytotoxin			
Cylindrospermopsin	Blocks protein synthesis; substantial cumulative toxicity	*Cylindrospermopsis raciborskii*	200 µg/Kg/d

Table 3. Examples of cyanobacterial toxins and their acute toxicity modified from [33]

5. The application of microbial source tracking in water

5.1. Microbial source tracking in water

Microbial source tracking (MST) is the use of microbial markers (including bacteriophage, bacteria, viruses, protozoans, etc.) to determine the source of fecal pollution present in an aquatic system (i.e., human, pet, livestock). Some refer to the scientific discipline as bacterial source tracking (BST), but this term is older and has given way to MST, acknowledging the wider array of information that can be gleaned from targets other than bacteria. The MST field continues to advance rapidly and much that has been published ten years ago is now obsolete and so it is generally better to refer to studies conducted since 2005 for the most updated information. For preliminary information on the evolution of the MST field, the reader should refer to the US Environmental Protection Agency's MST Guide (EPA 2005), article by [62]. MST methods were once categorized into two types — library dependent approaches and library independent approaches. Library dependent methods relied on matching the known source "fingerprints" or isolate patterns generated by molecular or phenotypic methods to unknowns collected from the environment. The most commonly used library dependent methods include, antibiotic resistant analysis (ARA), carbon utilization profiles (CUP), pulse field gel electrophoresis (PFGE), restriction- or amplified- fragment length polymorphism (RFLP or AFLP), random amplified polymorphic DNA (RAPD), repetitive extragenic palindromic PCR (rep-PCR) and ribotyping.

Library independent approaches are based upon detection (often presence/absence) or quantification of specific markers of fecal contamination, and are not dependent upon any sort of classification into categories. Library independent methods include F-specific and somatic coli phage serotyping and genotyping, and host-specific and marker-specific PCR and qPCR for a wide range of targets (see Table 3). Many older approaches utilized conventional PCR, but those have generally given way to the increased use of qPCR. Conventional PCR results are typically reported as presence/absence, or at best by using a serial dilution technique, can produce semi-quantitative information [63]. As opposed to conventional PCR, which is limited to a presence/absence result, qPCR provides for quantification over a wide dynamic range. The platforms currently being used for qPCR include hardware with proven rapid cycling, sensitive optics, and multiplex capabilities.

5.2. Bacteroides based microbial source tracking

One of the most fruitful organisms of focus in MST has been the Bacteroidales family. The Bacteroidales family has yielded a range of quantitative markers for human, dog, cow, horse, and other types of fecal contamination (e.g. 23, 71) The Bacteroides genus was suggested as early as 1967 as an alternative indicator due to the fact that *Bacteroides* spp. levels present in human and animal intestinal tracts are about 10 to 100 times above the other FIB [14, 16]. The high degree of difficulty for cultivation of *Bacteroides* spp. reduced its use in the late decades of the 20th century, but the development of molecular tools allowed the detection and quantification of members of the family Bacteroidales, without cultivation [64].

Consequently, there has been a dramatic increase in publications detailing development of qPCR assays for the quantification of Bacteroidales in recreational waters as indicators of specific types of fecal contamination [14,65]. While several studies have demonstrated the value of the family Bacteroidales, some have also highlighted problems associated with cross-reactivity for certain markers, making testing and validation of assays particularly important [64, 65]. One of the most important advancements has been the development of the HF183 human specific markers that has been shown in a range of studies to be specific and relatively sensitive [66,67,68]. Acknowledging that no one marker can be used everywhere and in all circumstances, the HF183 marker has been atypical is that it has been successfully utilized more than any other single marker in application for detection and quantification of human fecal contamination in recreational waters. Recent work [69] using the HF183 marker has shown that data generated can be suitable for providing stakeholders and decision markers with valuable information for remediation projects [31].

Recently, another qPCR-based method has been developed by [70] to quantify fecal *Bacteroides* spp. The assay was specifically designed to select for the *Bacteroides* species that are most human-associated, but it also amplifies Bacteroides spp. from animals. The focus of the assay is on a cohort of *Bacteroides* species that are most prominent in the human gut. This method has been successfully used as a screening tool, with more specific methods used to quantify specific types of fecal contamination [71]. Another successful approach is reported by [71], where the authors developed an array of fecal *Bacteroides* based markers for a range of types of fecal contamination. They reported a quantitative universal *Bacteroidales* assay called BacUni-UCD that detected universal fecal *Bacteroidales*in all of the test samples (n=73) including human, cat, seagull, horse, dog and cow feces. The human assay BacHum-UCD presented by [68], successfully discriminated fully between human and cow stool samples and had slight cross-reactivity with dog stool, however, the method is highly quantitative when used in combination with human specific approaches outlined by [68]. All of the wastewater samples tested during the [68], study were positive for the BacHum-UCD marker, showing it to be 100 percent sensitive for human source identification. In 2009, Ahmed and co-workers [73] published a test of the host specificity of the five published sewage-associated *Bacteroides* markers. They tested 186 fecal samples from 11 animal species including humans. All of the human fecal samples were positive for all five markers indicating 100 percent sensitivity of the above referenced markers of [71], [68] and [72]. The HF183 marker has been found to have the capability to differentiate between human and animal feces with 99 percent accuracy [73]. It has been used in several mid-Atlantic stormwater studies to indicate the presence of human fecal contamination [74].

5.3. Virus based methods

Pathogen-based research for MST has focused almost solely upon pathogens that are carried via the fecal-oral route. The work that has been accomplished regarding enteric viruses has relied on the fact that high concentrations of these viral pathogens can be found in human sewage. High concentrations are a vital characteristic, given the limitations with concentration and isolation of pathogens in wastewater and complex aquatic matrices. In 2005, in a review written by Fong and Lipp [75], the authors recount the major developments in the past decades, including applications of human and animal viruses to source tracking. The authors detail the disadvantages and advantages of virus-based detection and quantification methods. In particular, they present a summary of information of the myriad of animal viruses applied to source tracking studies for human fecal contamination, namely human enteroviruses (echovirus and Coxsackie virus for example) and adenoviruses (Ad5, 7, 40, and 41). The review is completed by iterating the need for strong study designs with the use of virus detection and quantification methods for microbial source tracking applications, due to the limitations with the ability to detect viral targets, which can often occur in aquatic systems at low concentrations and high patchiness.

Rajal and co-workers [76] presented work on the quantification of human and animal pathogenic viruses in real world water samples. They used a known sample volume and tracer addition of surrogate viruses to calculate the limit of detection. This work was one of the first to detail carefully the quantification and recovery in a tightly designed study. For any type of human pathogenic virus quantification, recovery estimations are vital. This can typically be done through the use of spiked samples or by adding surrogate DNA or RNA to the sample, or with the use of a competitive internal positive control (CIPC). In 2006, Gregory et al.[78] reported the first use of a CIPC for full quantification of RNA viruses for water quality applications. The point of this internal control was to permit assessment of the efficiency of the combined reverse-transcriptase and PCR steps to enterovirus quantification and to provide a scenario for CIPC design approaches for other virus types. Similar approaches were developed by [79] and have resulted in successful quantification of important human pathogenic enteroviruses in recent years.

While not all viruses will be infectious in a water sample, accurate quantification of viruses in a water sample along with flow measurements can permit loading estimates of potentially pathogenic material, specific to source, thereby permitting partitioning specific sources of fecal contamination in complex environments. Also, the accurate quantification of the human pathogenic viruses is an important parameter in the successful determination of the potential public health risk and can be a key component to Quantitative Microbial Risk Assessment (QMRA).

Recently, human polyomaviruses (HPyV) have emerged as a useful human-specific viral marker. There has been a series of studies conducted that report the presence and quantification of these viruses in raw sewage samples and/or environmental waters in sites ranging from Florida [82], California [82, 83], Australia [66], Spain [84], Germany [85] and Japan [86]. These viruses are double-stranded DNA viruses frequently isolated from urine, and in some cases feces, of both healthy and immunocompromised individuals. In a recent study conducted by [87], both nested PCR and qPCR were used to assess the presence of HPyV in raw and treated sewage in Rio de Janeiro, Brazil. The study interestingly demonstrated the clustering of different types of the HPyV coming from distinct African and European lineages. While this study was focused on samples from a sewage treatment plant, it demonstrates the specificity of information that can be gleaned from viral pathogen-based investigations.

A very recent publication by [87] has demonstrated the use of a range of MST approaches at a beach in southern California that has been the site of a range of water quality investigations (e.g. [43, 88]). Doheny State Beach, in Orange County, California, USA is a high priority, intensely utilized beach location that has suffered from a range of beach water quality problems, In the past decade, for example, the beach has been given a poor grade several times on the "Heal the Bay Beach Report Card" (www.healthebay.org). This study demonstrated the utility of the HF183 human specific marker, adenovirus, HPyV, and *Methanobrevibacter smithii* as markers of fecal contamination. They found distinct correlations among the highly human specific molecular markers, i.e. adenovirus at Doheny State Beach was correlated to both HPyVs and the HF183 molecular marker. This study dramatically presents the benefits of application of a range of microbial source tracking approaches (i.e. the use of the toolbox approach), as opposed to a single marker. It also demonstrates the complexity of real-world water quality investigations and the importance of conducting studies over appropriate time and sampling scales.

6. Conclusions

Over the past decades, the use of culture-based FIB based quantification approaches has dominated the management of microbial contaminants for drinking water, recreational water, and shellfish harvesting water quality. While these approaches are useful and have resulted in a dramatic improvement in water quality management over that observed prior to the 1950's, there is an array of new molecular approaches that can be used to bolster water quality management and the process of public notification (i.e. water advisories, beach posting and closings, and shellfish harvesting water closures).

Here, highlights have been presented of three areas for immediate advancement of the science. Rapid qPCR based methods are poised for future use to improve the accuracy of beach water quality postings and beach closures. Molecular assays specific for toxin producing strains of cyanobacteria, and other HAB species can be utilized to improve the management of our vital raw surface waters. Microbial source tracking techniques are poised to improve all facets of water quality management; from site specific investigations of the predominant fecal sources to assessment of the efficiency of wastewater treatment systems.

This is an exciting time for researchers given the fast evolution of molecular approaches. The design of appropriate methods, that are versatile, fast and reliable, can permit information to be transmitted to managers in shorter periods of time. The molecular methods will only improve in their cost effectiveness and user-friendliness in the next few years. Finally, the successful application of predictive multivariate models and formalization of QMRA approaches will improve dramatically with the infusion of the scale appropriate molecular information on microbial contaminants, sources of contamination and rates of discharge and loading.

Author details

Sandra Mesquita[1] and Rachel T. Noble[2]

1 Center Centre for Marine Sciences from Algarve (CCMAR), Campus de Gambelas, Portugal

2 Institute of Marine Sciences, University of Chapel Hill, North Carolina, Morehead City, USA

References

[1] Tallon P, Magajna B, Lofranco C and Leung KT. Microbial indicators of fecal contamination in water: a current perspective. Water, Air, and Soil Pollution, 2005; 166:139–166

[2] Bláha L, Babica P, Blahoslav. Toxins produced in cyanobacterial water blooms – toxicity and risks Interdisc Toxicol, 2009; 2(2): 36–41

[3] Standard methods for the examination of water and wastewater. Greenberg, A.E, Clesceri, L.S. and Eaton, A.D. (Eds). 18th edition; 1992;

[4] Balarajan R, Raleigh VS, Yuen P, Wheeler D, Machin D, Cartwright R.(1991) Health risks associated with bathing in seawater. British Medical Journal, 1991; 303: 1444

[5] Haile RW, Witte JS, Gold M, Cressey R, McGee C, Millikan RC, Glasser A, Harawa N, Ervin C, Harmon P, Harper J, Dermand J, Alamillo J, Barrett K, Nides M, Wang GY. The health effects of swimming in ocean water contaminated by storm drain runoff. Epidemiology, 1999; 10: 355-363

[6] Dufour AP, Ballentine P. Ambient water quality criteria for bacteria. Technical Report EPAA440/5-84-002. U.S. Environmental Protection Agency, Washington DC, 18, 1986

[7] Izbicki JA, Swarzenski PW, Reich CD, Carole Rollins C and Holden, PA. Sources of fecal indicator bacteria in urban streams and ocean beaches, Santa Barbara, California. Annals of Environmental Science, 2009; 3:139-178

[8] World Health Organization. Report of WHO Scientific Group of human viruses in water, wastewater and soil. Technical Report Series, no. 639. World Health Organization, Geneva, Switzerland, 1979

[9] Ikner LA, Gerba CP and Bright KR,. Concentration and Recovery of Viruses from Water: A Comprehensive Review on Food Environmental Virology, 2012;4:41–67

[10] Cao Y, John F Griffith JF, Samuel Dorevitch S and Stephen B Weisberg SB. Effectiveness of qPCR permutations, internal controls and dilution as means for minimizing

the impact of inhibition while measuring Enterococcus in environmental waters. Journal of Applied Microbiology 2012; 113(1):66-75

[11] World Health Organization. Report of WHO Guidelines for drinking water quality, I-recommendations. World Health Organization, Geneva, Switzerland, 2004;

[12] Figueras MJ and Borrego JJ.. New Perspectives in Monitoring Drinking Water Microbial Quality. Int. J. Environ. Res. Public Health, 2010; 7:4179-4202

[13] U.S. ENVIRONMENTAL PROTECTION AGENCY. Ambient Water Quality Criteria for Bacteria. EPA-440/5-84-002, U.S. Environmental Protection Agency, Washington, D.C., 1986

[14] Griffith JF, Cao Y, McGee CD, Stephen B. and Weisberg B. Evaluation of rapid methods and novel indicators for assessing microbiological beach water quality. Water Research, 2009; 43:4900 – 4907

[15] Harmsen, H J. M, Raangs GC, He T, Degener JE, and Welling GW. Extensive set of 16S rRNA-based probes for detection of bacteria in human feces. Applied and Environmental Microbiology, 2002; 68:2982–2990

[16] Leila Kahlisch l, Henne K, and Gröbe, L.. Assessing the Viability of Bacterial Species in Drinking Water by Combined Cellular and Molecular Analyses. Microb Ecol 2012; 63:383–397

[17] Eichler S, Christen R, Höltje C, Westphal P, Bötel J, BrettarI,Mehling A. and Höfle MG. Composition and dynamics of bacterial communities of a drinking water supply system as assessed by RNA- and DNA-based 16S rRNA gene fingerprinting. Applied and Environmental Microbiology, 2006; 72:1858–1872

[18] Paavola, J., Institutions and environmental governance: a reconceptualization. Ecological Economics 2007; 6, 1:93–103

[19] Orin C. Shanks, O C, Kelty,CA, Sivaganesan, M, Varma M and Richard A. Haugland, RA Quantitative PCR for Genetic Markers of Human Fecal Pollution, Applied and Environmental Microbiology, 2009;.75, 17:5507–5513

[20] U.S. ENVIRONMENTAL PROTECTION AGENCY, Ambient Water Quality Criteria for Bacteria. EPA-440/5-84-002, U.S. Environmental Protection Agency, Washington, D.C., 1986

[21] Dafour, AP. Health Effects Criteria for Fresh Recreational Waters. EPA-600/1-84-004, U.S. Environmental Protection Agency, Research Triangle Park, N.C., 1984

[22] Davies, CM, Long, JA, Donald, M and N. J. Ashbolt, NJ, Survival of fecal microorganisms in marine and freshwater sediments. Applied and Environmental Microbiology 1995; 61:1888–1896

[23] Layton, A., McKay L,Williams D, Garrett V, Gentry, R and Sayler. G. Development of Bacteroides 16S rRNA gene TaqMan-based real-time PCR assays for estimation of to-

tal, human, and bovine fecal pollution in water. Applied Environ Microbiology 2006; 72:4214–4224

[24] Matsuki T, Watanabe K, Fujimoto J, Takada T, and Tanaka, R. Use of 16S rRNA gene-targeted group-specific primers for real-time PCR analysis of predominant bacteria in human feces. Applied and Environmental Microbiology, 2004; 70:7220–7228

[25] Boehm, AB, Ashbolt, NJ. A sea change ahead for recreational water quality criteria. Journal of Water Health 2009; 7, 1: 9-20

[26] Gentry, R W, Layton AC, McKay LD, McCarthy JF, Williams DE, Koirala SR, and Sayler GS. Efficacy of Bacteroides measurements for reducing the statistical uncertainty associated with hydrologic flow and fecal loads in a mixed use watershed. Journal of Environmental Quality, 2007; 36:1324–1330

[27] Figueras MJ, Robertson W, Pike E and Borrego JJ. Sanitary inspection and microbiological water quality. In: Monitoring Bathing Waters. A Practical Guide to the Design and Implementation of Assessments and Monitoring Programs; Bartram J. and Rees G (Eds), E & FN Spon: London, UK, 2000; pp. 113-126

[28] Fielda KG and Samadpourb M. Fecal source tracking, the indicator paradigm, and managing water quality. Water Research, 2007; 41: 3517-3538

[29] Environmental Health Directorate – Health Protection Branch. Recreational Water Quality-Published by Authority of the Minister of National Health and Welfare, 1978;

[30] Kornacki JL and Johnson JL. Enterobacteriaceae, Coliforms and *Escherichia coli* as quality and safety indicators'. In: F. Downs (Eds), Compendium of Methods for the Microbiological Examination of Foods, APHA. Washington DC, 2001;

[31] World Health Organization. Guidelines for drinking-water quality, 3rd ed., vol. 1. World Health Organization, Geneva, Switzerland; 2008

[32] WHO, Guidelines for Drinking-Water Quality, 2nd edition Geneva, 1993; ISBN 92 4 154460(Albinana-Gimenez et al. 2009), Germany (Hamza et al. 2009) and Japan (Haramoto et al. 2010)

[33] Chorus I and Bartram J (Eds.). Toxic cyanobacteria in water: a guide to their public health consequences, monitoring and management. E&FN Spon, ISBN 0-419-23930-8, London, 1999;

[34] Izbicki JA, Swarzenski PW, Reich CD, Carole Rollins C and Holden, PA. Sources of fecal indicator bacteria in urban streams and ocean beaches, Santa Barbara, California. Annals of Environmental Science, 2009; 3:139-178

[35] Whitman RL, Przzybyla-Kelly K,Shivley DA and Byappanhali MN. Incidence of the enterococcal surface protein (esp) gene in human and animal fecal sources. Environ SciTechnol 2007; 41(17):6090–6095

[36] Dafour, AP. Health Effects Criteria for Fresh Recreational Waters. EPA-600/1-84-004, U.S. Environmental Protection Agency, Research Triangle Park, N.C., 1984

[37] Leecaster MK and Weisberg SB. Effect of Sampling Frequency on Shoreline Microbiology Assessments. Marine Pollution Bulletin, 2003;42(11): 1150-1154.

[38] Pisciotta J, et al. The role of neutral lipid nanospheres in *Plasmodium falciparum* crystallization. Biochem J, 2007, 402:197–204.

[39] Haugland RA, Siefring SC, Wymer LJ, Brenner KP and Dufour AP. Comparison of Enterococcus density measurements by quantitative polymerase chain reaction and membrane filter culture analysis at two freshwater recreational beaches. Water Research, 2005; 39:559-568.

[40] Griffith JF and Weisberg SB. Challenges in Implementing New Technology for Beach Water Quality Monitoring: Lessons from a California Demonstration Project. Marine Technology Society Journal 2011; 45:65–73

[41] Noble RT, Blackwood AD, Griffith JF, McGee CD and Weisberg SB. Comparison of Rapid Quantitative PCR-Based and Conventional Culture-Based Methods for enumeration of *Enterococcus* spp. and *Escherichia coli* in Recreational Waters. Applied and Environmental Microbiology, 2010; 76(22):7437-7443

[42] Wade TJ, Calderon RL, Brenner KP, Sams E, Beach M, Haugland R, Wymer L and A. P. Dufour AP. High Sensitivity of Children to Swimming-Associated Gastrointestinal Illness Results Using a Rapid Assay of Recreational Water Quality. Epidemiology, 2008; 19: 375-383

[43] Colford JM, Schiff K, Griffith JF, Yau V, Arnold BF, Wright C, Gruber J, Wade T, Burns S, Hayes S, McGee C, Gold M, Noble RT and Weisberg, SB. Using rapid indicators for Enterococcus to assess the risk of illness after exposure to urban runoff contaminated marine water. Water Research, 2012; 1-11.

[44] Scott TM, Jenkins TM,, Lukasik J and Rose JB. Potential use of a host-associated molecular in *Enterococcus faecium* as an index of human fecal pollution. Environmental Science and Technology, 2005; 39 (1):283–287.

[45] McQuaig SM, Scott TM, Harwood VJ, Farrah SR and Lukasik JO. Detection of human-derived fecal pollution in environmental waters by use of a PCR-based human polyomavirus assay. Applied and Environmental Microbiology, 2006; 72(12):7567–7574

[46] Lavender JS and Kinzelman JL. A cross comparison of QPCR to agar-based or defined substrate test methods for the determination of *Escherichia coli* and enterococci in municipal water quality monitoring programs. Water Research 2009; 43(19): 4967-4979

[47] Tsai YL and. Olson BH. Rapid method for separation of bacterial DNA from humic substances in sediments for polymerase chain reaction. Applied and Environmental Microbiology, 1992; 479(58):2292-2295

[48] De Boer SH, L.J. Ward LJ, X. Li X and S. Chittaranjan. S. Attenuation of PCR in- hibition in the 411 presence of plant compounds by addition of BLOTTO. Nucleic Acids Research, 1995; 23:2567-2568

[49] Kreader CA. Relief of amplification inhibition in PCR with bovine serum albumin or T4gene 32 protein. Applied and Environmental Microbiology, 1996; 62:1102-1106

[50] Watson RJ and B. Blackwell. Purification and characterization of a common soil component which inhibits polymerase chain reaction. Canadian Journal of Microbiology, 2000; 46:633-642

[51] Thurman EM, Aiken GR, Ewald M, Fischer WR, Forstner U, Hack AH, Mantoura RFC, Parsons JW, Pocklington R, Stevenson FJ, Swift RS and SzpakowskaB.Isolation of soil and aquatic humic substances. pp. 31-43 in: F.H. Frimmel and R.F. Christman (ed.), Humic Substances and Their Role in the Environment. John Wiley and Sons, Ltd. New York, NY, 1988

[52] Cao Y, Griffith JF and Weisberg SB. Evaluation of optical brightener photo decay characteristics for detection of human fecal contamination. Water Research, 2009; 49 (8):2273-227

[53] Bushon RN, Brady AM, Likirdopulos CA and Cireddu JV. Rapid detection of *Escherichia coli* and enterococci in recreational water using an immunomagnetic separation/ adenosine triphosphate technique Journal of Applied Microbiology, 2009;106: 432–441

[54] Lee CM, Griffith JF, Kaiser W and Jay, JA. Covalently linked immune magnetic separation/ adenosine triphosphate technique (Cov-IMS/ATP) enables rapid, in-field detection and quantification of *Escherichia coli* and Enterococcus spp. in freshwater and marine environments. Journal of Applied Microbiology, 2009; 109:324–333

[55] Lee JY. and Deininger RA. Detection of *E-coli* in beach water within 1 hour using immunomagnetic separation and ATP bioluminescence. Luminescence 2004; 19:31–36

[56] Wiegand C and Pflugmacher S. Ecotoxicological effects of selected cyanobacterial metabolites a short review. Toxicology and Applied Pharmacology, Vol. 203, (March 2005); pp. 201-218, ISSN 0041-008X

[57] Al-Tebrineh J, Leanne A. Pearson LA, Serhat A. Yasar SA and Brett A. Neilan BA. A multiplex qPCR targeting hepato- and neurotoxigenic cyanobacteria of global significance. Harmful Algae, 2012; 15:19–25

[58] Sivonen K and Jones G. Cyanobacterial toxins. In: Toxic cyanobacteria in water: a guide to public health significance, monitoring and management. Chorus I and Bar-

tram (Eds).The World Health Organization, ISBN 0-419-23930-8, E and FN Spon, London, UK, 41-111pp, 1999

[59] Conradie KR and Barnard S. The dynamics of toxic *Microcystis* strains and microcystin production in two hypertrofic South African reservoirs. Harmful Algae 2012; in press

[60] Henderson, Chips M, Cornwell, Hitchins P, Holden B, Hurley S, Parsons SA, Wetherill A and Jefferson B. Experiences of algae in UK waters: a treatment perspective. Water and Environment Journal, 2008; 22:184–192

[61] Pinckney JL, Richardson TL, Millie DF, Paerl HW. Application of photopigment biomarkers for quantifying microalgal community composition and in situ growth rates. Organic Geochemistry, 2001; 32: 585–595

[62] Simpson JM, Santo Domingo JW and Reasoner DJ. Microbial source tracking: state of the science. Environmental Science and Technology 2002; 36:5279–5288

[63] Noble RT and J. A. Fuhrman, JA. Enteroviruses detected in the coastal waters of Santa Monica Bay, CA: Low correlation to bacterial indicators. Hydrobiologia, 2001;460:175-184

[64] Gentry, R W, Layton AC, McKay LD, McCarthy JF, Williams DE, Koirala SR, and Sayler GS. Efficacy of Bacteroides measurements for reducing the statistical uncertainty associated with hydrologic flow and fecal loads in a mixed use watershed. Journal of Environmental Quality, 2007; 36:1324–1330

[65] Boehm, AB, Ashbolt, NJ. A sea change ahead for recreational water quality criteria. Journal of Water Health, 2009; 71: 9-20

[66] Ahmed W, Stewart J, Powell D and Gardner T. Evaluation of Bacteroides markers for the detection of human faecal pollution. Letters in Applied Microbiology, 2008; 46(2): 237-242

[67] Ahmed W, Goonetilleke A, Powell D, Chauhan K and Gardner T. Comparison of molecular markers to detect fresh sewage in environmental waters. Water Research, 2009;43:4908-4917

[68] Seurinck S, Defoirdt T, Verstraete W and Siciliano SD.Detection and quantification of the human-specific HF183 Bacteroides 16S rRNA genetic marker with real-time PCR for assessment of human faecal pollution in freshwater. Environmental Microbiology, 2005; 7(2):249-59

[69] Mieszkin S, Furet JP, Corthier G, and M. Gourmelon M. Estimation of pig fecal contamination in a river catchment by real-time PCR using two pig-specific Bacteroidales 16S rRNA genetic markers. Applied and Environmental Microbiology 2009; 75:3045–3054

[70] Converse RR, Blackwood AD, Kirs M, Griffith JF and Noble RT, Rapid QPCR-based assay for fecal Bacteroides spp. as a tool for assessing fecal contamination in recreational waters. Water Resources, 2009; 43:4828–4837

[71] Kildare BJ, Leutenegger CM, McSwain SM, Bambic DG, Rajal VB and Wuertz S. 16S rRNA-based assays for quantitative detection of universal, human-, cow-, and dog-specific fecal Bacteroidales: A Bayesian approach. Water Research, 2007; 41:3701-3715

[72] Bernhard A and Field, KG. Identification of nonpoint sources of fecal pollution in coastal waters by using host-specific 16S ribosomal DNA genetic markers from fecal anaerobes. Applied and Environmental Microbiology, 2000; 66:1587–1594

[73] Ahmed W, Goonetilleke A, Powell D, Chauhan K and Gardner T. Comparison of molecular markers to detect fresh sewage in environmental waters. Water research, 2009; 43:4908-4917

[74] Parker JK, McIntyre D and Noble RT. Characterizing fecal contamination in stormwater runoff in coastal North Carolina, USA. Water Resources, 2010; 44(14):4186-4194

[75] Fong TT and Lipp EK. Enteric viruses of humans and animals in aquatic environments: health risks, detection, and potential water quality assessment tools. Micro Mol Biol Rev, 2005; 69(2):357-371

[76] Rajal, McSwain BS, Thompson DE, Leutenegger CM, Wuertz, S. Molecular quantitative analysis of human viruses in California stormwater. Water Research, 2007; 41(19):4287-4298

[77] Wong M, Kumar L, Jenkins TM, Xagoraraki I, Phanikumar MS and Rose JB, (2009) Evaluation of public health risks at recreational beaches in Lake Michigan via detection of enteric viruses and a human-specific bacteriological marker. Water re- sources, 2009; 43:1137-1149

[78] Gregory JB, Litaker RW and Noble RT. Rapid One-Step Quantitative Reverse Transcriptase PCR Assay with competitive internal positive control for detection of Enteroviruses in environmental samples. Applied and Environmental Microbiology, 2006; 72(6):3960-3967

[79] Fuhrman JA, Liang X and Noble RT. Rapid Detection of Enteroviruses in Small Volumes of Natural Waters by Real-Time Quantitative Reverse Transcriptase PCR Applied and Environmental Microbiology, 2005; 71(8): 4523–4530

[80] Baums L, Goodwin KD, Kiesling T, Wanless D, Diaz MR, Fell, JW. Luminex detection of fecal indicators in river samples, marine recreational water, and beach sand. Marine Pollution Bulletin, 2007; 54 (5):521-536

[81] Carson CA, Christiansen JM, Yampara-Iquise H, Benson VW, Baffaut C, Davis JV, Broz RR, Kurtz WB, Rogers WM and Fales WH. Specificity of *Bacteriodes thetaiotaomicron* marker for human feces. Applied and Environmental Microbiology, 2005; 71(8):4945-4949

[82] McQuaigSM , Scott TM , Lukasik JO , Paul JH , Harwood VJ . Quantification of human polyomaviruses JC virus and BK virus by TaqMan quantitative PCR and comparison to other water quality indicators in water and fecal samples .Applied and Environmental Microbiology 2009; 75: 3379-3388

[83] Rafiquea and Jiang, CS. Genetic diversity of human polyomavirus JCPvV in Southern California wastewater. Journal of Water Health, 2009; 6:533-538

[84] Albina-Gimenez N, Clemente-Casares P, Calgua B, Huguet, JM, Courtois S and Girones R. Comparison of methods for concentrating human adenoviruses, polyomavirus JC and noroviruses in source waters and drinking water using quantitative PCR. J Virol Methods, 2009; 158: 104-109

[85] Hamza IA, Jurzik L, Stang K, Uberla K and Wilhem M. Detection of human viruses of a densly-populated area in Germany using virus adsorption method optimized for PCR analyses. Water Resources, 2009; 43:2657-2668

[86] Haramoto E, Kitajima M, Katayama H and Ohgaki S. Real-time pCR detection of adenoviruses, polyomaviruses and torque viruses in river water in Japan. Water resources, 2010; 44:1747-1752

[87] Fumian TM, Guimarães FR, Pereira Vaz BJ, da Silva MT, Muylaert FF, Bofill-Mas S, Gironés R, Leite JP and Miagostovich MP. Molecular detection, quantification and characterization of human polyomavirus JC from waste water in Rio De Janeiro, Brazil, 2010; 8(3):438-45

[88] McQuaig S, Griffith J and Harwood VJ. The Association of Fecal Indicator Bacteria with Human Viruses and Microbial Source Tracking Markers at Coastal Beaches Impacted by Nonpoint Source Pollution Applied and Environmental Microbiology AEM.00024-12; published ahead of print 6 July 2012

Chapter 3

GIS-Assisted Modelling of Soil Erosion in a South African Catchment: Evaluating the USLE and SLEMSA Approach

G. D. Breetzke, E. Koomen and W. R. S. Critchley

Additional information is available at the end of the chapter

http://dx.doi.org/10.5772/52314

1. Introduction

Soil erosion – the detachment and transportation of particles from soil aggregates by erosive agents (Stocking, 1984) – is regarded as one of South Africa's most significant environmental problems (Meadows, 2003). In South Africa, roughly 6 million households derive all or some of their income from agriculture (South African Department of Agriculture, 2007). Roughly 25 % of the population is directly dependant on agriculture, an activity that utilises about 80% of the total surface area of the country (Lutchmaih, 1999). The implications of high soil erosion rates are reflected in agricultural costs as well as social welfare costs where the decline in soil productivity causes the migration of the workers to urban areas. For a developing country, such as South Africa, these social impacts further burden the national economy.

Crucial in combating the scourge of soil erosion in South Africa is to estimate amounts and rates of soil loss in the country at various levels of scale. This will facilitate the initiation of regional land-use and management planning strategies, and the application of appropriate conservation and management practices at various scales of development. Studies on soil erosion in South Africa have been summarised by Garland, Hoffman & Todd (1999); annual soil loss estimates for the whole country range from 363 million tonnes (Midgley, 1952), to 233 million tonnes (Schwartz & Pullen, 1966) and to 100-150 million tonnes (Rooseboom, 1976). These overall national figures are based on the sediment yield of main rivers in South Africa. While indicating the importance of soil erosion on a national level these figures are unhelpful in drafting local or regional plans to combat erosion. Conservation strategies are conventionally planned on the scale of river catchments; at this scale the complete erosion process is included while it is still possible to spatially pinpoint actual control measures.

Geographical Information Systems (GIS) provide a much-favoured tool in regional soil erosion studies in South Africa (Le Roux et al., 2007). Such tools facilitate the upscaling of plot-scale soil loss predictions to a catchment or bigger scale.

In this paper we apply GIS technology to estimate soil loss rates per land use type in a quaternary catchment using two common approaches that generate rapid soil erosion assessment results at a low cost: the Universal Soil Loss Equation (USLE) and the Soil Loss Estimator for Southern Africa (SLEMSA). The objective of our research is to critically compare these popular approaches and discuss potential ways of improving their application. This paper starts with a concise introduction to the study site and then discusses how soil erosion is described in the USLE and SLEMSA approaches. Specific attention is paid to the way the various constituting factors are made operational. Subsequently the resulting soli loss estimates are described and their relation with the underlying factors is analysed. In a final concluding section the main results are highlighted and their implications for other applications of these approaches are discussed.

2. Study site and methodology

As study site we selected a catchment in the KwaZulu-Natal province of South Africa. The catchment is situated between latitudes 29º 30' 36" S and 29º 52' 48" S and longitudes 29º 8' 24" E and 29º 5' 24" E and has a surface area of approximately 341 km^2 (Figure 1). The altitude of the catchment ranges from 1160 m a.s.l from the Wagendrift Dam, at the outlet of the catchment, to 2080 m at the Giant's Castle nature reserve at the western corner of the catchment. The topography is characterised by deeply incised valleys and steep slopes mainly covered by grassland. The catchment is located in a sub-humid environment and receives an annual average rainfall of 932mm. Rainfall is concentrated in the summer months (November - March) with the winter months (May – August) receiving as little as 10mm of rainfall per month. The most notable water body in the catchment is the Bushman's River that drains into the Wagendrift Dam at the outlet of the catchment.

Grassland covers over 80% of the catchment with the remainder consisting mainly of forest plantations and thicket and scrubland. A small percentage (4,3%) of the catchment consists of small-scale subsistence, and large-scale commercial agricultural settlements. The commercial settlements comprise of six holdings in the catchment. The small-scale settlements, where subsistence agriculture is practiced, are numerous and sporadic. The commercial farmers focus mainly on dairy production, while subsistence agriculture, practiced by approximately 5000 families, is based mainly on the food grains of maize and sorghum. The geology of the catchment is characterised mainly by dark-grey (often carbonaceous) shale, siltstone and fine and medium to coarse-grained sandstone (Turner, 2000). There is a great diversity of fauna and flora in the catchment as well as several national parks, the most notable being the Wagendrift nature reserve and the Giant's Castle nature reserve.

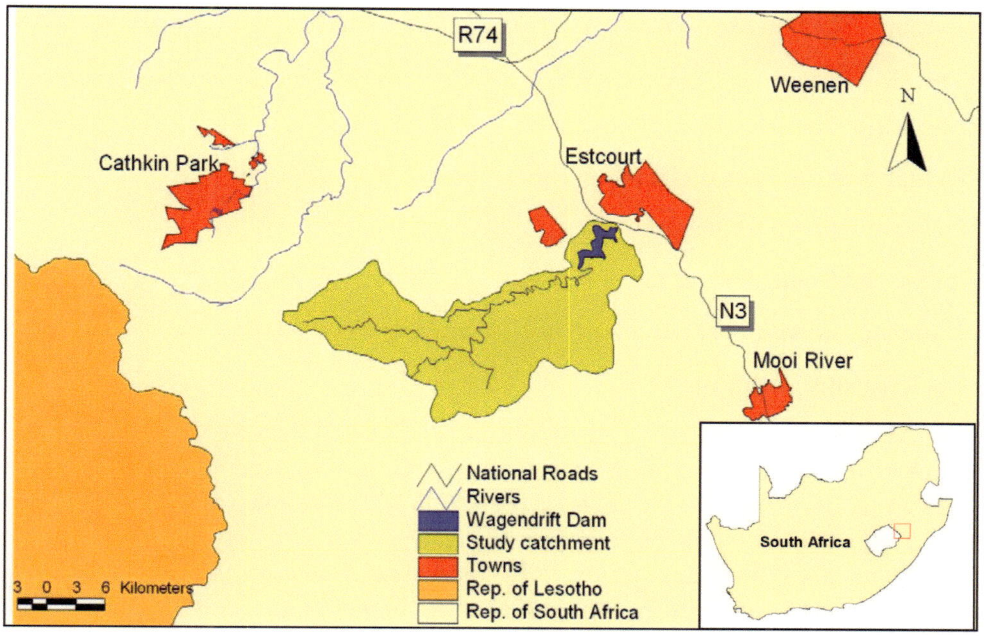

Figure 1. Location of the study catchment area in KwaZulu-Natal, South Africa

2.1. Governing equations

The USLE was developed in 1965 by the Agricultural Research Service (ARS) scientists Wischmeier and Smith to predict long time average soil losses in run-off from specific field areas in specified cropping and management systems (Wischmeier & Smith, 1978). The USLE disaggregates the erosion process into 6 factors that were each determined based on the analyses of more than 11 000 plot-years of research data from 47 locations in 24 states in the United States. Notwithstanding its initial north American focus, this approach or its revised successor (RUSLE, see Renard et al., 1991) has been applied in many studies around the world including a particularly interesting study that estimates sediment yield in the past 6000 years in the Meuse catchment area (Ward et al., 2009). The basic equation follows:

$$A = R \, x \, K \, x \, LS \, x \, C \, x \, P \tag{1}$$

Where:

A = Mean annual soil loss (t ha^{-1} yr^{-1})

R = Rainfall and runoff erosivity index (J mm.m^{-2} h^{-1})

K = Soil erodibility factor (t J^{-1} mm^{-1})

LS = Slope and length of slope factor

C = Cropping – Management factor

P = Erosion control factor practice

The SLEMSA model was developed by Elwell (1977) in Zimbabwe to estimate the long-term mean annual soil loss from sheet erosion on arable land (Bonda et al., 1999). SLEMSA was developed on the basis of the USLE and is an attempt to adapt the USLE model to an African environment. It is a relatively widely used soil loss model in African environments (Elwell & Stocking, 1982), and should be seen as a modelling technique or framework, rather than mechanistic descriptions of the erosion system (Smith, 1999). The SLEMSA model divides the soil erosion environment into four physical systems: crop, climate, soil and topography. The SLEMSA equation is represented schematically as follows:

$$Z = K \, x \, C \, x \, X \tag{2}$$

Where:

Z = Mean annual soil loss from the land (t ha^{-1} yr^{-1})

K = Erodibility factor (t ha^{-1} yr^{-1})

C = Crop factor

X = Topographic factor

2.2. Calculating the USLE factor values

GIS is used to calculate the individual USLE factor grids that, upon multiplication, provide the total potential soil loss within the catchment. A short description of the assumptions and calculations related to the creation of the factor grids is provided below. For a more comprehensive explanation of the methodology see Breetzke (2004).

2.2.1. Rainfall erosivity (R) factor

Cubic surface trend analysis was used to create a Mean Annual Precipitation (MAP) isohyetal grid of the site, based on an average of 30 years of annual rainfall data. The rainfall-erosivity grid was obtained by assigning a regional specific formula based on a rainfall-erosivity relationship developed by the Department of Agriculture and Water Supply (1984) to the MAP grid. The equation is based on computed erosion index values (EI_{30}) for a rainfall station located within the site and is shown below. The erosion index, EI_{30}, for a given storm is a product of the kinetic energy of the falling raindrops and its maximum 30-minute intensity (Engel, 2002):

$$R = 0.63P - 153.72 \tag{3}$$

Where:

P = mean annual precipitation grid (in mm)

2.2.2. Soil erodibility (K) factor

The erodibility factor was calculated according to the nomograph method outlined in Wischmeier & Smith (1978) and shown mathematically below. Basic data for estimating soil erodibility were obtained by collecting 120 samples from test sites representative of the major soil-mapping units in the catchment. The erodibility was calculated as:

$$K = \left(2.1 * 10^{-4} * \left(12 - OM\right) * M^{1.14} + 3.25\left(s - 2\right) + 2.5\left(p - 3\right)\right) / 759 \tag{4}$$

Where:

K = erodibility factor (in ton/MJ/mm)

OM = organic matter content (%)

M = texture product

s = structure class

p = permeability class.

A fine particle analysis was conducted to obtain the percentage sand, silt, clay and organic matter for each test site. These values were used to obtain a soil erodibility value per test site. The K factors generated for each test site were subsequently used as a variable for the erodibility grid map composed using the Inverse Distance Weighting (IDW) interpolator. This grid map was summarised to create a table containing mean K values per soil type in the catchment, and a grid created with the mean K values as the variable.

2.2.3. Topographic (LS) factor

The topographic factor consists of two sub-factors: a slope gradient factor and a slope length factor. A DEM was built through digitising the contours of a 1:50 000 topographic map of the study site. The slope length and slope gradient factors (shown below) were calculated using the filled DEM and entered into the equation below to produce the topographic factor grid, following:

$$LS = (L/22)^{0.5}\left(0.065 + 0.045S + 0.0065S^2\right) \tag{5}$$

Where:

$L = (x/22.13)^m$, in which x is length of slope (in m), m is 0.5 if the slope is >5 %, 0.4 if between 3 and 5 %, 0.3 if between 1 and 3 percent and 0.2 if below 1and L is the slope length factor;

$S = (0.43 + 0.30\ s + 0.043\ s^2)/6.613$, where s is the gradient (%), and S is the slope gradient factor.

2.2.4. Crop management (C) factor

Land use types in the site were assigned C factor values based on their percentage canopy cover, fall height and ground cover. These values are determined using Thompson's (1996) classification, aerial photo analysis, information from studies conducted within southern Africa on specific crops and land cover types, (e.g. McPhee & Smithen, 1984) and field observation of the catchment. In this way mimicking similar research (e.g. Donald, 1997), in determining appropriate C factor values for a South African catchment in which little local data is available.

2.2.5. Erosion control practice (P) factor

Information on the support practices or P factor values in the site (e.g. contour intervals, terracing, burning) was collected through field observation. Field examination of the land cover-mapping units revealed the only form of erosion control being practiced in the site was contour tillage on land under temporary cultivation. According to McPhee and Smithen (1984) a support practice factor value of 0.6 should be assigned to land cover under this control practice and the remainder of the site is assigned the P factor value of 1, indicating no physical evidence of erosion control in these areas.

2.3. Calculating the SLEMSA factor values

GIS is used to calculate the individual SLEMSA factor grids that, upon multiplication, provide the total potential soil loss within the catchment. A short description of the assumptions and calculations related to the creation of the factor grids is provided below. For a more comprehensive explanation of the methodology see Breetzke (2004).

2.3.1. Erodibility (K) factor

The erodibility factor of SLEMSA is determined using the exponential relationship (Morgan, 1995):

$$lnK = blnE + a \tag{6}$$

Where:

E represents the kinetic energy of raindrops as they strike the soil or vegetation, in J/m^2 (Schultze, 1979); and

a and b are functions of the soil erodibility factor (F).

Schultze (1979) calculated a rainfall intensity and kinetic energy equation for the region which is shown below and used to calculate the kinetic energy of the raindrops, E:

$$E = 15{,}16MAP - 1517.67\ J.\,m^{-2}annum^{-1} \tag{7}$$

Where:

MAP equals mean annual precipitation (in mm)

The erodibility (F) of the soil is governed by its soil texture and soil type. Using the results of the particle size analysis and governed by the United States Department of Agriculture (USDA) textural triangle, the texture of 120 soil samples at test sites was determined. An individual soil erodibility value (F) was subsequently assigned to each test site according to the specifications provided by Elwell (1978). An erodibility value per test site was derived and used as a variable for the erodibility grid map composed using the IDW interpolator. This grid map was summarised to create a table containing the mean K values per soil type in the catchment, and a grid created with the mean K values as the variable.

2.3.2. Slope length (X) factor

The slope length factor consists of two sub-factors: a slope gradient factor and a slope length factor. The slope length and slope gradient factors are calculated using the filled DEM and entered into the equation below to produce the slope length factor grid.

$$X = \sqrt{\left(L * \left(0.76 + 0.53 * S + 0.076 * S^2\right) / 25.65\right)} \tag{8}$$

Where:

X = topographic ratio

L = slope length, in metres (m)

S = slope steepness, in percent (%)

2.3.3. Crop (C) factor

The crop factor (C) is based on a Zimbabwean model originally developed for grassland by Elwell and Stocking (1976). A summary of the factor is shown below:

$$C = e^{(-0.06i)} \text{when } i < 50\% \tag{9}$$

and

$$C = (2{,}3 - 0{,}01i)/30 \text{ when } i > 50\% \tag{10}$$

Where:

C = the ratio of soil loss from a crop having an interception value of i, compared to the soil loss from bare fallow;

i = percentage rainfall energy intercepted by the crop

The average percentage cover values for the land cover types were adapted from Schultze's (1979) index. Validation of these observations was provided through research done by Elwell (1977) and Edwards (1967).

3. Soil loss estimates

The erosion potential according to the USLE and SLEMSA models is shown in Figures 2 and 3 respectively. Soil loss rates are classified into five categories ranging from very low, where soil loss rates range between 0–1 $t^{-1}.ha^{-1}.yr^{-1}$, to very high where soil loss rates exceed 25 $t^{-1}.ha^{-1}.yr^{-1}$. Table 1 indicates the soil loss rate per land use type in the site. Soil loss rates were classified according to land use types as this allows for an effective subdivision of each soil model's input parameters thus providing useful insight into the components that contribute to the calculated soil loss rates. This knowledge can further aid planners in determining effective soil conservation strategies at the regional level. The basic conclusion drawn of Table 1 is that USLE and SLEMSA provide an average rate per hectare ($t.ha^{-1}.yr^{-1}$) of differing magnitude. Large differences are indicated per land use type where SLEMSA greatly exceeds the USLE models' mean annual rates on the of unmanaged grassland, thicket and scrubland, and indigenous forest land use types, while on the cultivated land use types, the USLE mean annual rates provide higher estimates.

http://dx.doi.org/10.5772/52314

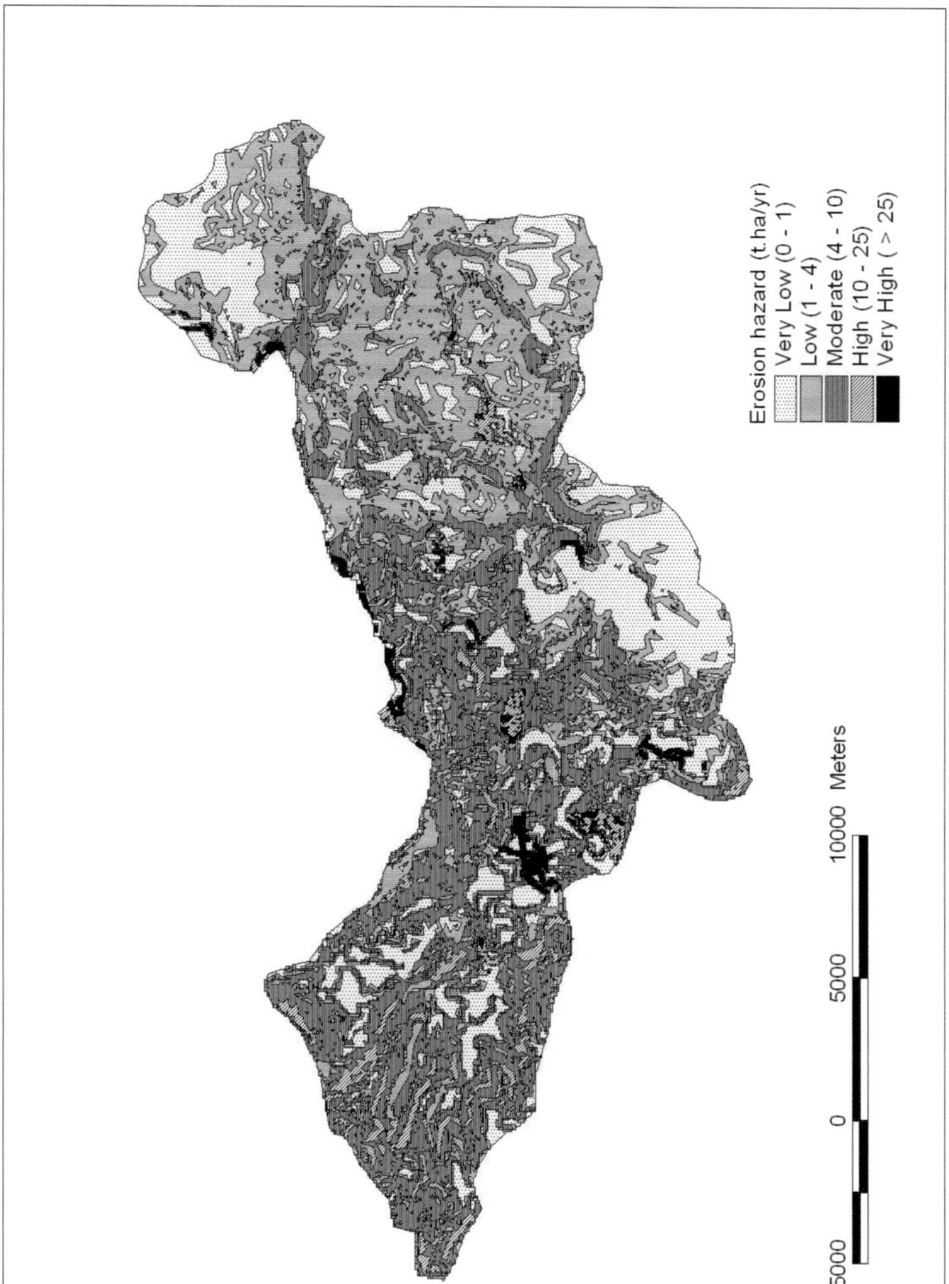

Figure 2. USLE soil erosion hazard in the study catchment

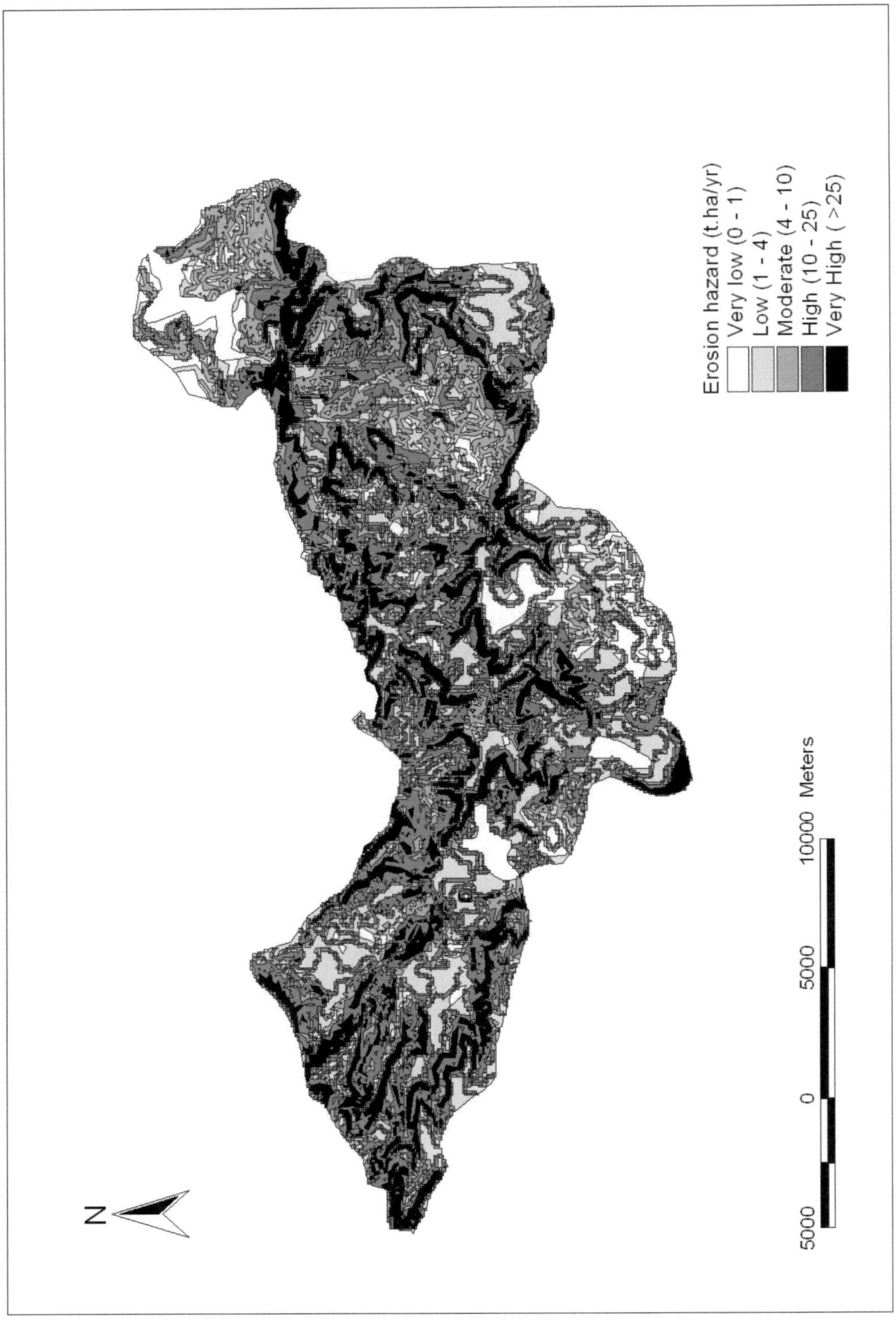

Figure 3. SLEMSA soil erosion hazard in the study catchment

http://dx.doi.org/10.5772/52314

Land-use type[1]	Area (km²)	Coverage (%)	Soil loss approach (t.ha⁻¹.yr⁻¹)	
			USLE	SLEMSA
Grassland (unmanaged)[2]	278.2	81.5	4.1	15.6
Forest plantations[3]	18.7	5.5	0.6	2.8
Thicket and scrubland[4]	16.4	4.8	1.5	13.6
Cultivated: TCD [5]	6.3	1.9	8.5	2.4
Cultivated: TSD[6]	5.8	1.7	16.2	5.4
Residential land[7]	5.6	1.6	8.6	3.1
Waterbodies[8]	5.6	1.6	-	-
Cultivated: TCI[9]	2.2	0.7	13.3	2.8
Indigenous Forest[10]	1.5	0.4	2.1	30.8
Grassland (managed)[11]	1.1	0.3	0.6	4.3
Average rate per hectare (t.ha⁻¹.yr⁻¹)			4.11	13.88

Table 1. Estimated mean annual soil loss per land use type

Notes:

1. Land cover types are in accordance with the land classification of the CSIR's – Satellite Application Centre of South Africa (CSIR-SAC, 2001).

2. Essentially indigenous species, growing under natural or semi-natural conditions.

3. All areas of systematically planted, man-managed tree resources, composed of primarily exotic species (including hybrids).

4. Areas of densely interlaced trees and shrub species (often forming an impenetrable community).

5. TCD: temporary commercial dryland: Large, uniform, mechanised, well-managed field units under temporary crops with lack of major irrigation schemes.

6. TSD: temporary subsistence dryland: Small field units in close proximity to rural population centres. Typically crops produced for individual or local (i.e. village) markets. Low level of mechanisation. Low-level mechanisation.

7. Areas in which people reside on a permanent or near-permanent basis.

8. Areas of (generally permanent) *open water*.

9. TCI: temporary commercial irrigated: Large, uniform, mechanised, well managed field units under temporary crops using major irrigation schemes.

10. Planted grassland, containing either indigenous or exotic species, growing under man-managed conditions for grazing, hay or turf production or recreation (e.g. golf courses)

3.1. USLE soil loss per factor

Overlaying the USLE erosion map on the grids of rainfall erosivity, soil erodibility, slope and length of slope, and crop management factors provides information on those land use types associated with high soil loss rates. A statistical comparison of mean factor rates of USLE and predicted soil loss values is provided in Table 2.

In general, the mean soil loss rate per land use type correlates most significantly with the mean crop management factor values. The cultivated land use types in the catchment had typically the highest mean crop management factor values which are indicative of the low percentage canopy cover, fall height and ground cover of the cultivated land. Such low ground cover values are typically found on cultivated land in South Africa where the national ground cover of cultivated land rarely exceeds 40% (Thompson, 1996). The canopy cover in the catchment, although affected by the current growth phase of the crop, rarely exceeded 50% as the crops are temporary (i.e. annuals) and are harvested at the completion of the growing season but remain idle until replanted, therefore prone to severe erosion in heavy rainfall events. The fallow period of crops by subsistence farmers in rural KwaZulu-Natal is extensive and coincides with rainfall peaks, particularly in the summer months. The canopy cover of forestland in the catchment (i.e. indigenous forest and forest plantations) on the other hand is continuous. It comprises mostly of evergreen trees beneath which the vegetation is multi-layered (Bredenkamp *et al.*, 1996).

The emphasis that USLE places on the ground and canopy cover explains the low soil loss rates of forestland in the USLE approach when compared to the SLEMSA estimates that rather consider the percentage rainfall energy that is intercepted by the crop than on the percentage canopy cover, fall height and ground cover.

Land use type	Mean factor rate					Soil loss
	R	*K*	*LS*	*C*	*P*	(t. h^{-1} yr^{-1})
Grassland (unmanaged)	629	0.016	10.5	0.040	1.0	4.1
Forest plantations	650	0.015	6.1	0.006	0.6	0.6
Thicket and scrubland	586	0.017	16.6	0.008	0.6	1.5
Cultivated: TCD	557	0.016	4.0	0.421	1.0	8.5
Cultivated: TSD	739	0.017	8.4	0.170	1.0	16.2
Built-up – residential	753	0.014	6.3	0.130	1.0	8.7
Water bodies	-	-	-	-	-	-
Cultivated: TCI	493	0.019	4.1	0.673		13.3
Indigenous forest	764	0.015	25.0	0.006	1.0	2.1
Grassland (managed)	479	0.016	7.2	0.008	1.0	0.6

Table 2. Comparison of mean factor rates of USLE and mean predicted soil loss values per land use type

http://dx.doi.org/10.5772/52314

3.2. SLEMSA soil loss per factor

Overlaying the SLEMSA erosion map on the grids of topographic, erodibility and crop factors provide information on those land use types associated with a high soil loss rates. A statistical comparison of the mean factor rates of SLEMSA and predicted soil loss values is shown in Table 3. In general, the mean soil loss rate per land use type correlates most significantly with the topographic factor values and indicates the strong influence of the topographic factor plays in determining soil loss estimations in SLEMSA. In general, erosion rates were low on gradual slopes (e.g. cultivated land and forest plantations) with mean soil loss rates less than 2,5 t. h^{-1} yr^{-1} predicted. Erosion rates were typically high on steep slopes (e.g. thicket and scrubland, and indigenous forest) with mean soil loss rates greater than 30 t. h^{-1} yr^{-1} predicted.

Land use type	Mean factor rate			Soil loss
	K	*C*	*X*	(t. h^{-1} yr^{-1})
Grassland (unmanaged)	29.8	0.12	4.3	15.6
Forest plantations	30.8	0.05	1.9	2.8
Thicket and scrubland	28.8	0.05	9.2	13.6
Cultivated: TCD	29.6	0.06	1.3	2.4
Cultivated: TSD	30.7	0.05	3.2	5.4
Built-up – residential	32.8	0.04	2.2	3.1
Water bodies	-	-	-	-
Cultivated: TCI	27.8	0.06	1.4	2.8
Indigenous forest	29.9	0.05	20.8	30.8
Grassland (managed)	32.9	0.06	2.1	4.3

Table 3. Comparison of mean factor rates of SLEMSA and mean predicted soil loss values per land use type

A further subdivision of the topographic factor of SLEMSA into slope degree, slope gradient (S) and slope length (L), Table 4, indicates that slope gradient, in particular, is the over-riding factor in explaining the high soil loss rates attributed to certain land use types. The mean slope length attributed to each land use type remains relatively consistent throughout the site, while the slope gradient is highest on those land use types with similarly high-predicted erosion rates. A point supported by Hudson (1987) who found that soil loss estimations using SLEMSA in mountainous terrain in South Africa were very sensitive to variations in slope steepness, while le Roux et al (2004) established that SLEMSA predicts excessive high soil losses on steep slopes and regions with high rainfall, while conducting a catchment scale studying using SLEMSA in Mauritius. In general, the slope length of the land use types are too small to bring about a concentrated flow and the effect of the slope gradients on the input parameters within SLEMSA is significant as it is the predominant factor that influences the erosion rates.

Land use type	Mean factor rate		
	Slope (°)	*Slope factor S*	*Slope length L*
Grassland (unmanaged)	0.07	137.5	3.0
Forest plantations	0.07	23.4	2.7
Thicket and scrubland	0.07	75.3	3.3
Cultivated: TCD	0.07	20.9	2.2
Cultivated: TSD	0.07	53.3	2.7
Built-up – residential	0.07	26.0	2.6
Water bodies	-	-	-
Cultivated: TCI	0.07	17.1	2.5
Indigenous forest	0.30	130.9	3.3
Grassland (managed)	0.07	6.1	2.8

Table 4. Comparison of slope related mean factor rates and mean predicted SLEMSA soil loss values per land use type

4. Discussion and conclusion

An accurate validation of the soil loss rates obtained is challenging, as there is a dearth of empirical investigations covering soil loss in South Africa and no calculated soil loss data from run-off plots in the catchment. It is beyond the scope of the study to develop a set of field data to assess the accuracy of each model but rather the focus is confined to qualitatively contrasting the soil loss rates obtained for each model and elaborating on causal influences within each model that play a significant role in eliciting the varying soil loss rates obtained.

The strikingly different results illustrated pose the question whether or not the use of USLE or SLEMSA for erosion modelling at a catchment scale is valid. The selection of both soil loss models to a mountainous quaternary catchment in South Africa must raise questions of applicability. Numerous studies have been conducted investigating the use of USLE in South African conditions, most notably, Donald (1997), McPhee & Smithen (1984) and Crosby, McPhee & Smithen (1983), these researchers propose that USLE could be applied to South African conditions provided input data for local conditions could be developed. Site-specific correct parameters however, have not been determined for both models in South Africa and neither model has been comprehensively tested and calibrated to determine its practicality in a South African environment. Yet the majority of soil erosion prediction research conducted in South Africa has been done using the USLE, Revised Universal Soil Loss Equation (RUSLE) and SLEMSA models (Smith, 1999). The USLE (e.g. Smith *et al.*, 2000) and SLEMSA (e.g. Schulze, 1979; Hudson, 1987) have, however, been applied on catchment scales elsewhere and these studies demonstrate that the models are capable of adequately predicting soil loss under different land use, despite being applied to conditions beyond the original database (Le Roux *et al.*, 2004).

The spatial scales with which these models have been applied in practice are not the spatial scale for which they have been conceived. The mismatch between the small spatial and temporal scales of data collection and model conceptualisation, and the large spatial and temporal scales of most intended uses of models (Renschler & Harbor, 2002) is a major challenge in soil erosion modelling, which has become even more important with the increasing use of models linked to GIS. The problem with spatial scale and erosion modelling is two fold – on the one hand by estimating potential soil loss at a catchment scale the spatial error in the application is propagated. Jetten et al (1999) states the reason being that at a catchment scale the input maps are often created from a limited amount of field data and with a lot of assumptions and therefore highly subjective; there are also many methods of interpolation that are equally valid but give different results. He concludes that all these problems mean that there is a greater opportunity for concatenation and amplification of any errors and uncertainties in the input data within the model. On the other hand however GIS is able to model catchment-scale applications and treat heterogeneous catchments of varying size to produce regional results for a catchment-scale conservation strategy.

4.1. Future developments

The USLE and SLEMSA soil loss models were used to estimate soil loss rates in a quaternary catchment in the KwaZulu-Natal province of South Africa. The mean annual soil loss is estimated approximately at 4.11 $t.ha^{-1}.yr^{-1}$ by the USLE and 13.88 $t.ha^{-1}.yr^{-1}$ by SLEMSA. The SLEMSA rates greatly exceed the USLE rates on the unmanaged grassland, thicket and scrubland, and indigenous forest land use types, while on the cultivated land use types, the USLE mean annual rates provide higher estimates. Overlaying the USLE and SLEMSA erosion maps on the respective factor grids provided an insight into factors that played a significant role in eliciting the varying soil loss rates obtained. For the USLE, the crop management factor provided the most significant influence in determining high soil loss rates, whereas in SLEMSA the topographic factor was the predominant factor that influenced the erosion rates per land use type. Our analysis shows that SLEMSA is very sensitive to variations in slope steepness whereas the effects of crop and canopy cover within USLE are the strongest determinants of high erosion potential.

The USLE and SLEMSA factor calculations and resulting local soil loss estimates need to be validated by measuring erosion from run-off plots or applying a correction for inter-catchment deposition by means of a Sediment Delivery Ratio (SDR). To date such validation work is lacking (Le Roux et al., 2007). Developing effective regional values for land use types and soils within each soil erosion assessment should occur as uncertainty regarding the allocation of crop factor and soil erodibility values within a study can have a significant impact on the results produced, particularly within a South African context. The focus on the variable results obtained should however be shifted towards what models are best suited for each spatial application, the problem for developing countries, where data is scarce and unreliable, is that they do not have a choice in the selection of a model for determining erosion potential simply because of a lack of data resources. Both the methodology and results obtained through the paper pose the question whether or not such a study can stand up to scientific scrutiny, the answer is provided in the lack of the realistic alternatives for soil loss estimation in South Africa. For the foreseeable future, the USLE and SLEMSA as well as the methodology employed still have a role to play.

Acknowledgements

Pam. Porter and Nicolene Fourie at the Chief Directorate of Surveys and Mapping in Mowbray, Cape Town provided electronic data. Hendrik Smith at the ARC - Institute for Soil, Climate and Water provided documents required for application in the study. Merete Styczen at the Danish Hydrology Institute (DHI) provided the ArcView software extension, SEAGIS that led to the development of the study. Use of laboratory facilities at the University of Pretoria.

Author details

G. D. Breetzke[1], E. Koomen[2] and W. R. S. Critchley[3]

1 Department Of Geography, University Of Canterbury, New Zealand

2 Department of Spatial Economics, VU University Amsterdam, the Netherlands

3 Resource Development Unit, Centre For International Cooperation, VU University Amsterdam, the Netherlands

References

[1] Bonda, F., Mlava, J., Mughogho, M., and Mwafongo, K., 1999. Recommendations for future research to support erosion hazard assessment in Malawi. Malawi Environmental Monitoring Programme Report. University of Arizona, USA.

[2] Bredenkamp, G., Granger, J.E., Hoffman, T., Low, B., Lubke, R.A. Mckenzie, B., Rebelo, G. and Van Rooyen, N., 1996. Vegetation of South Africa, Lesotho and Swaziland. Department of Environmental Affairs and Tourism, Pretoria.

[3] Breetzke, G. D., 2004. A critique of soil erosion modelling at a catchment scale using GIS. Unpublished MSc thesis, Vrije Universiteit Amsterdam, The Netherlands.

[4] CSIR-SAC, 2001. Illustrated field guide. Council For Scientific And Industrial Research - Satellite Applications Centre (CSIR-SAC), Pretoria.

[5] Crosby, C.T., Mcphee, P.J. and Smithen, A.A., 1983. Role of soil loss equations in estimating sediment production, In: H. Maaren (ed), Proceedings of the Workshop on the effect of rural land use and catchment management on water resources, TR 113, Pretoria, 188-213.

[6] Department of Agricultural and Technical Services, 1976. Soil Loss Estimator for Southern Africa, Natal Agricultural Research Bulletin 7.

[7] Department of Agriculture and Water Supply, 1984. National Soil Conservation Manual, Chapter 6: Predicting rainfall erosion losses, 1-12.

[8] Donald, P.D., 1997. GIS modelling of erosion and sediment yield in a semi-arid environment. MSc. Thesis, University of the Witwatersrand.

[9] Edwards, D., 1967. A plant ecological survey of the Tugela River Basin. Natal Town and Regional Planning Commission, Pietermaritzburg.

[10] Elwell, H.A. and Stocking, M.A., 1976. Vegetal cover to estimate soil erosion hazard in Rhodesia. Geoderma (Amsterdam), 15, 61-70.

[11] Elwell, H.A. and Stocking, M.A., 1982. Developing a simple yet practical method of soil loss estimation. Tropical Agriculture (Trinidad), 59, 43-48.

[12] Elwell, H.A., 1977. A soil loss estimation system for southern Africa. Rhodesian Department of Conservation and Extension, Research Bulletin, No. 22.

[13] Elwell, H.A., 1978. Soil loss estimation; compiled works of the Rhodesian multi-disciplinary team on soil loss estimation. Institute of Agricultural Engineering, Borrowdale, Salisbury.

[14] Engel, B., 2002. Appendix A: Estimating Soil Loss with the USLE. College of engineering, Purdue University, West Lafayette, IN

[15] Garland, G., Hoffman, T. and Todd, S., 1999. Soil Degradation. In: Hoffman, T., Todd, S., Ntshona, Z. and Turner, S., 1999, Land Degradation in South Africa. National Botanical Institute of South Africa.

[16] Hudson, C.A., 1987. A regional application of SLEMSA in the Cathedral Peak area of the Drakensberg. Unpublished MSc. Thesis, University of Cape Town, South Africa.

[17] Jetten, V., De Roo, A. and Favis-Mortlock, D., 1999. Evaluation of field-scale and catchment-scale soil erosion models. Catena, 37, 521–541.

[18] Le Roux, J.J., Sumner, P.D. and Rughooputh, S.D.D.V., 2005. Erosion modelling and soil loss prediction under changing land use for a catchment on Mauritius. South African Geographical Journal, 87, 129-140.

[19] Le Roux, J.J., Newby, T.S., Sumner, P.D., 2007. Monitoring soil erosion in South Africa at a regional scale: review and recommendations. South African Journal of Science, 103 (7-8), 329-335.

[20] Lutchmiah, J., 1999. Soil erosion in the central midlands of KwaZulu-Natal: A comparative study. The South African Geographical Journal, 81, 143-147.

[21] Mcphee, P.J. and Smithen, A.A., 1984. Application of the USLE in the Republic of South Africa. Agricultural Engineering in South Africa. 18 (1), 5-13.

[22] Meadows, M.E., 2003. Soil erosion in the Swartland, Western Cape Province, South Africa: Implications of past and present policy and practice. Environmental Science & Policy, 6, 17–28.

[23] Midgley, D.C., 1952. A preliminary survey of the surface water resources of the Union of South Africa. PhD thesis. University of KwaZulu-Natal, Pietermaritzburg.

[24] Morgan, R.P.C., 1995. Soil Erosion and Conservation. (2nd ed.), (UK: Longman).

[25] Renard, K.G., Foster, G.R., Weesies, G.A. and Porter, J.P., 1991. RUSLE-revised universal soil loss equation. Journal of Soil and Water Conservation 46, 30-33.

[26] Renschler, C.S. and Harbor, J., 2002. Soil erosion assessment tools from point to regional scales – the role of geomorphologists in land management research and implementation. Geomorphology, 47, 189-209.

[27] Rooseboom, A., 1976. Reservoir sediment deposition rates. In 2nd International Congress on Large Dams, Mexico.

[28] Schultze, R.E., 1979. Soil loss in the Key Area of the Drakensberg – a regional application of the 'Soil Loss Estimation Model for Southern Africa' (SLEMSA). In: Hydrology and Water Resources of the Drakensberg, Pp. 149-167, Natal Town and Regional Planning Commission, Pietermaritzburg, South Africa.

[29] Schwartz, H.I. and Pullen, R.A., 1966. A guide to the estimation of sediment yield in South Africa. Civil engineering in South Africa, pp. 343-346.

[30] Smith, H.J., 1999. Application of Empirical Soil Loss Models in southern Africa: a review. South African Journal of Plant and Soil, 16, 158-164.

[31] Smith, H.J., Van Zyl, A.J., Claasens, A.S., Schoeman, J.L. and Laker, M.C., 2000. Soil loss modelling in the Lesotho Highlands Water Project catchment areas, South African Geographical Journal, 82, 64-69.

[32] South African Department Of Agriculture, 2007, Strategic plan DoA 2007; Part 2: Sectoral overview and performance. Department of Agriculture, Pretoria.

[33] Stocking, M.A., 1984. Erosion and Soil Productivity: a Review. Soil Conservation Programme, Land and Water Development Division, Rome.

[34] Thompson, M., 1996. A standard land-cover classification for remote-sensing applications in South Africa. South African Journal of Science, 92, 34-42.

[35] Turner, D.P., 2000, Soils of KwaZulu-Natal and Mpumalanga: recognition of natural soil bodies. PhD thesis. University of Pretoria, Pretoria.

[36] Ward, P.J., Van Balen, R.T., Verstraeten, G., Renssen, H. and Vandenberghe, J., 2009. The impact of land use and climate change on late Holocene and future suspended sediment yield of the Meuse catchment. Geomorphology 103, 389–400.

[37] Wischmeier, W.H. and Smith, D.D., 1978. Predicting rainfall erosion losses. Agricultural Handbook 537. United States Department of Agriculture. Agricultural Research Service, Washington, DC.

Chapter 4

Estuarine Fisheries Community-Level Response to Freshwater Inflows

James M. Tolan

Additional information is available at the end of the chapter

http://dx.doi.org/10.5772/52313

1. Introduction

Every estuary needs freshwater inflow (FWI) to maintain proper salinity regimes, nutrient loading, and sediment inputs to support its geographically unique levels of biological productivity [1-4]. Watershed elevations and soil types determine surface and groundwater flows into estuaries, and these flows have source, timing, and velocity components that can be significantly affected by anthropogenic alterations at the landscape level. It is estimated that approximately 60% of the global storage of freshwater is now contained behind reservoirs and dams [5] and 77% of the total water discharge from 139 of the largest river systems in the northern hemisphere are either strongly or moderately affected by dams, interbasin transfers, and surface water withdrawals [6]. Hydrologic modifications of estuarine watersheds influence wetland and open-water salinity patterns, nutrients, sediment fertility, bottom topography, dissolved oxygen, and concentrations of xenobiotics [7]. Because demand for freshwater is only expected to increase as population continues to grow [8], it is incumbent upon resource managers to examine the environmental effects and biological consequences of hydrologic alterations within coastal ecosystems [9-11].

Resource-based approaches seek to link freshwater inflows to a number of fishery species generally considered valuable by society [12]. The optimization model utilized by Powell and Matsumoto [13] uses a series of relationships between monthly inflows and the catch of a number of commercially and recreationally important finfish (red drum *Sciaenops ocellatus*, black drum *Pogonias cromis*, spotted seatrout *Cynoscion nebulosus*, southern flounder *Paralichthys lethostigma*), crustaceans (blue crab *Callinectes sapidus*, white shrimp *Litopenaeus setiferus*, brown shrimp *Farfantepenaeus aztecus*, pink shrimp *F. duorarum*) and mollusks (eastern oyster *Crassostrea virginica*) to arrive at a set of targeted monthly freshwater inflows to maintain healthy ecological conditions in estuaries. The goal of this method, which jointly considers the salinity tolerance range of each of the target organisms and limits the inflow volume solution by imposing numerous process constraints (such as fishery biomass and harvest ratios; monthly, bi-monthly, and yearly freshwater volumes; upper and lower bounds for salinity; nutrient and sediment loading; see [13]), is to estimate the minimum amounts of FWI needed to maintain historical fisheries production. Although the inflow-harvest equations were originally based on fishery-dependent commercial catch records, recent modeling efforts have incorporated a greater proportion of fishery-independent data sources [14].

A problem with the resource-based approach is that it focuses on adults, which are harvestable. Although these are estuarine dependent species, the adults are widely distributed along salinity gradients [15]. A number of transient taxa (sensu, facultative estuarine-dependent, see [16]) which recruit from offshore spawning areas are known to have size-specific use patterns within shallow habitats of the oligohaline-to-freshwater portions of estuaries [17-20]. Inflows, especially those large pulses associated with flooding events, can displace seaward the boundary between the brackish and freshwater interface, on a kilometers-to-estuary wide scale. Taxa with specific nursery habitat requirements could therefore be restricted from ingress into portions of the estuary, potentially altering an important habitat for juvenile nekton.

The goal of this study was to expand the focus of interest of the resource-based approach beyond the limited number of fisheries target species and to include juvenile stages of fisheries species to examine the functional role of FWI in shaping the total nekton assemblage structure in estuaries. The approach was to perform an analysis of a long-term, state agency, bag seine monitoring program. Bag seine samples are fishery independent and contain juvenile stages of fishery species. There are three FWI gradients examined; within estuaries from river to sea, among estuaries along a climatic gradient, and over time as changes in freshwater inflows, in the form of flood pulses and drought events, dramatically alter the salinity structure of the estuary.

2. Materials and methods

2.1. Study area

Texas' coastline extends along 600 km of open Gulf of Mexico shoreline and contains 3,420 kilometers of bay-estuary-lagoon shoreline. This is a biologically rich and ecologically diverse region of the state, supporting more than 247,576 hectares of fresh, brackish, and salt marshes. Within the state, over 305,600 km of rivers and streams coalesce into 15 major river systems, and these rivers empty into seven major estuaries (Figure 1). All seven estuaries have similar geomorphic structure and physiography, yet each is quite diverse hydrologically. This is primarily due to a climatic gradient influencing freshwater inflows. This gradient of decreasing rainfall from northeast to southwest is one of the most distinctive features of the coastline (Table 1). Along this gradient, rainfall decreases by a factor of two, yet inflow decreases by almost two orders of magnitude. The Laguna Madre, a hypersaline lagoon, has a negative inflow balance because this estuary lacks any major riverine inflow and evaporation normally exceeds precipitation. The net effect is a gradient of estuaries with similar physical characteristics but greatly differing salinity regimes.

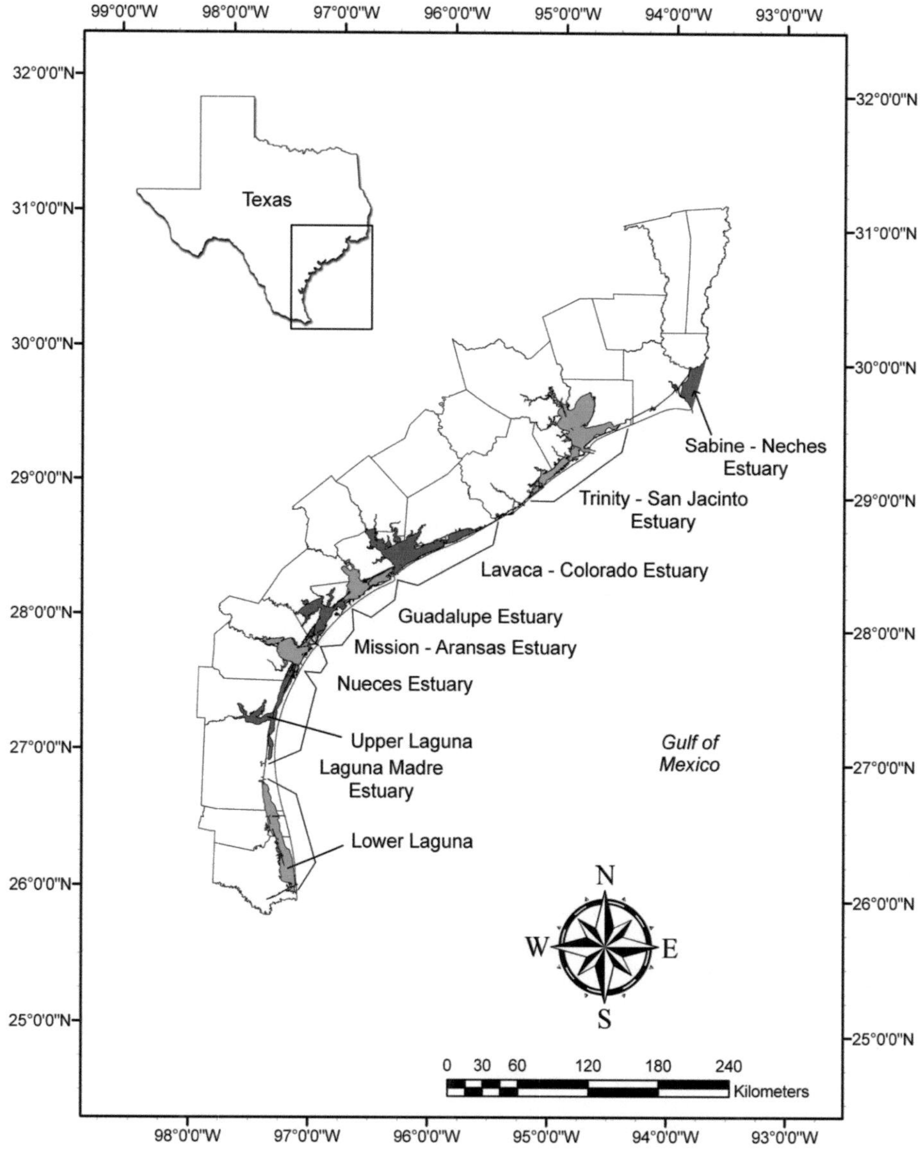

Figure 1. Map of Texas showing the location along the coast of each of the major estuarine systems.

Estuary	Drainage (km²)	Area (km²)	Rainfall (cm y⁻¹)	Inflow (10⁶ m³ y⁻¹)	Salinity (PSU)	Habitat Wetlands (km²)	Habitat SAV (km²)
Sabine-Neches	45 705	243	142	16 897	8	967	-
Trinity-San Jacinto	57 900	1399	112	14 000	16	1594	73
Lavaca-Colorado	111 890	1158	102	3801	18	348	28
Guadalupe	26 330	551	91	2664	16	271	65
Mission-Aransas	7860	453	81	265	19	393	85
Nueces	43 350	433	76	298	29	121	53
Laguna Madre	29 695	3658	69	-893	36	1825	773

Table 1. Climatic gradient in Texas estuaries, listed from north to south. Drainage basin size (Drainage), bay surface area (Area), and Habitat (SAV = submerged aquatic vegetation) characteristics from USEPA (1999). Average annual Inflow Balance, Rainfall, and Salinity characteristics (1941-1999) from the Texas Water Development Board; http://www.twdb.state.tx.us/data/bays_estuaries/bays_estuary_toc.asp

From the Louisiana border to the Trinity-San Jacinto estuary the coastline is characterized by marshy plains with low, narrow beach ridges, and from there to the border of Mexico the coastline is characterized by long barrier islands and large, shallow lagoons. Barrier islands are parallel to the mainland along the coast, and between the barrier islands and the mainland are lagoons. These lagoons are interrupted with drowned river valleys that form the bays and estuaries. Inlets through the barrier island connect the Gulf of Mexico to these lagoons, with each lagoon opening into a large primary bay. There is typically a constriction between the primary and secondary (and in some cases tertiary) bays. While ungauged coastal watershed runoff can locally influence estuarine salinity, most inflow into each bay is supplied primarily by just one or two gauged rivers draining hydrologically isolated watersheds.

2.2. Bag seine community structure data

Starting in January 1992 and continuing through the present, 20 replicate bag seine samples have been collected each month within each major estuary system along the Texas coast. Sampling locations are randomly selected from a grid system of one minute latitude and one minute longitude, with no selected grid sampled more than once per calendar month. For each sample, a bag seine (18.3 m X 1.8 m, 1.9 cm stretch nylon multifilament; central bag, 1.8 m wide, 1.3 cm stretch mesh) is pulled parallel to the shore for 15.2 m [21]. The surface area sampled is estimated using the distance pulled and length of extension of the bag seine. All fish and invertebrates collected in each sample are identified, enumerated, and measured. Total catch of each taxon is standardized and expressed as catch per hectare. Prior to each bag seine collection, surface salinity in Practical Salinity Scale were measured with either handheld Hydrolab or YSI multiprobes calibrated to the manufacturers specifications.

http://dx.doi.org/10.5772/52313

2.3. Data analysis

The experimental unit defined for this study was each bay system as a whole, as each of the major estuaries along the Texas coast can be defined by the underlying hydrologic gradient. To assemble the bag seine collections into a time series, catch data for each taxa from the monthly replicate samples were summed across estuaries individually, and reported as total catch per month. Salinity records were similarly transformed into a time series, although the replicate values were first averaged across each estuary and reported as mean salinity per month. For each estuary, a categorical 'Inflow Condition' variable was defined by evaluating the average salinity time series. Salinities above the 85th percentile were deemed indicative of 'Drought' conditions and values below the 15th percentile representative of high flow or 'Flood' conditions. Values between these two extremes were identified as 'Normal' flows. The Laguna Madre Estuary was further sub-divided into an Upper and Lower components, providing for eight estuaries under investigation (see Figure 1). This sub-division is based on a natural sand sheet or land bridge (the Land Cut) connecting Padre Island with the mainland [21]. The Gulf Intracoastal Waterway bisects this extensive sand flat, thus connecting the two lagoons via the Land Cut.

Community analyses were performed using Primer-E (Version 6.0) software [22]. A matrix of Bray-Curtis Distance similarities between each total catch per month sample was created. Catch data was initially transformed [Log_{10} + 1] to down-weight the most abundant taxa. Significant differences in rank similarities between groups of samples were then tested by Analysis of Similarity (ANOSIM). In the ANOSIM procedure, the probability of *a priori* groupings of samples is estimated by repeated permutations of the original data matrix. Values of the R statistic can range from -1 to 1, although R will usually fall between 0 and 1 with R values > 0.4 indicating higher degrees of discrimination among groups. The *a priori* factors tested with the ANOSIM procedure were the external factors associated with each sample (e.g., season of collection and inflow condition) within a common estuary. The entire collection was then merged across estuaries, and then tested for differences in community structure among estuaries. The SIMPER (SIMilarity PERcentages) routine was used to examine the contribution of individual species to the community structure seen among the *a priori* factors. Similarities among the samples are graphically represented with non-metric multidimensional scaling (MDS) ordinations [23]. Although outcomes of the ANOSIM are not dependant on MDS ordinations, the ordinations are presented here as they are a helpful way of visualizing patterns in the data. Stress values indicate how well the two-dimensional plot represents relationships among samples in the multidimensional space. Stress values < 0.15 indicate a good fit. MDS ordinations may be arbitrarily rotated so axes are not labeled.

3. Results

From 1992 through 2006, bag seine sampling resulted in 28,786 individual collections from the eight major estuarine systems. This time series of 180 months revealed dramatically fluctuating mean salinities throughout the study period (Table 2). Despite dramatic differences in total inflows across the coastal hydrologic gradient, temporal inflow patterns were generally similar across the coast. The timing of extended flooding conditions (i.e., on the order of 10 months during 1992 and again in 1997) or droughts (the majority of the calendar years of 2000 and 2001) were similar within each estuary (Figure 2).

	n	**Mean**	**Std. Dev.**	**Min.**	**Max.**
Estuary					
Sabine-Neches	3599	6.73	6.23	0.0	32.0
Trinity-San Jacinto	3597	16.52	9.21	0.0	41.0
Lavaca-Colorado	3599	18.47	9.57	0.0	40.0
Guadalupe	3594	16.62	11.36	0.0	45.0
Mission-Aransas	3600	18.13	9.70	0.0	41.0
Nueces	3600	28.66	7.42	0.0	59.0
Upper Laguna Madre	3599	35.44	10.64	0.0	78.0
Lower Laguna Madre	3598	32.04	7.98	0.0	64.0

Table 2. Salinity summary statistics by estuary for the study period January 1992 through December 2006.

The bag seines recorded 3,583,061 individuals from 387 unique taxa. Analysis of Similarity of the entire collection showed that in each estuary, community structure was significantly different across seasons (Table 3), and these seasonal differences were repeated annually across all inflow categories. The greatest disparity in community composition involved comparisons across opposite seasons (e.g., winter vs. summer, spring vs. fall), with significant pairwise comparison R values ranging from 0.609 – 0.971 (see Table 3). While seasonal differences in communities were quite evident, there appears to be little correspondence between community structure and synoptic-scale inflow events (Figure 3). This general disconnect between shallow water nekton assemblages and inflows was evident in every estuary along the Texas coast (Figure 4). The only estuaries to display significant community-level differences across the different inflow conditions were from opposite ends of the salinity spectrum. The Sabine-Neches estuary (mean salinity approximately 7) had significantly different community compositions during a drought relative to flood conditions ($R = 0.520$, $p < 0.001$). Greater abundances of white shrimp (7 fold increase), brown shrimp (17 fold increase), pinfish *Lagodon rhomboides* (6 fold increase), white mullet *Mugil curema* (12 fold increase), spotted seatrout (17 fold increase), and sheepshead minnow *Cyprinodon variegatus* (9 fold increase) were recorded during the periods of elevated salinities. The Lower Laguna Madre (mean salinity 32) also had significantly different community compositions during drought conditions (Drought vs. Normal comparison, $R = 0.253$, $p < 0.001$; Drought vs. Flood comparison, $R = 0.235$, $p < 0.001$), although the elevated salinities in this estuary during drought conditions (mean salinity > 40) led to lower abundances of some of these same taxa. Substantial decreases in brown shrimp (5 fold), white shrimp (15 fold), and Atlantic croaker *Micropogonias undulatus* (7 fold), as well as lower abundances of rainwater killifish *Lucania parva* (4 fold decrease) and red drum (2 fold decrease) were noted during extended low inflow conditions.

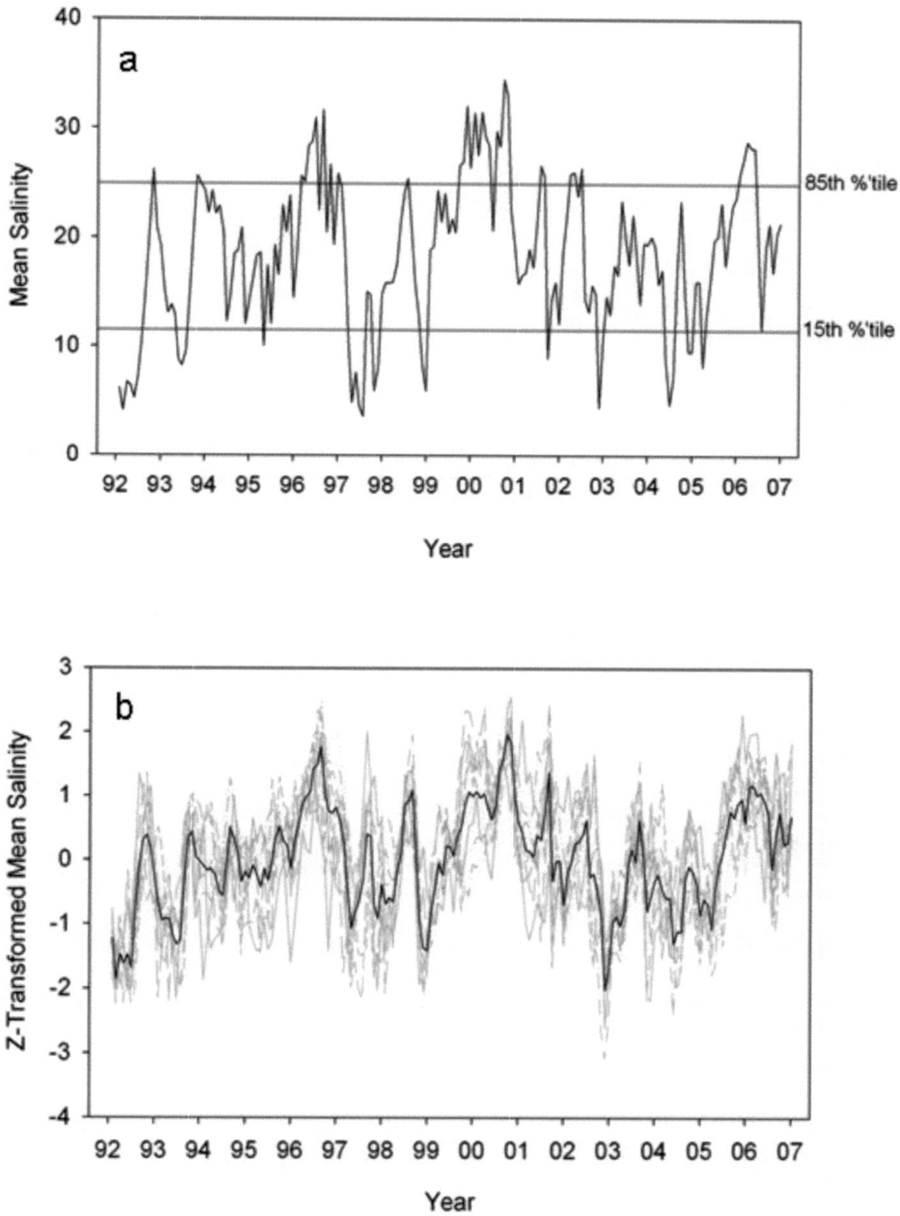

Figure 2. Estuarine-wide mean salinity time series during the study period of 1992 through 2006: (a) from a representative estuary (Lavaca-Colorado), and (b) from all eight major estuaries. Salinities in (b) are Z-transformed (not labeled individually for clarity), and a mean line added to aid in interpretation.

Estuary	Global R		Seasonal Pairwise Comparison R		
			Winter	Spring	Summer
		Spring	0.464	-	
Sabine-Neches	0.669***	Summer	**0.908**	0.566	-
		Fall	**0.836**	**0.834**	0.501
		Spring	0.628	-	
Trinity-San Jacinto	0.649***	Summer	**0.900**	0.553	-
		Fall	**0.815**	**0.786**	0.362
		Spring	0.576	-	
Lavaca-Colorado	0.695***	Summer	**0.971**	**0.697**	-
		Fall	**0.820**	**0.796**	0.391
		Spring	0.573	-	
Guadalupe	0.661***	Summer	**0.897**	0.660	-
		Fall	**0.756**	**0.786**	0.411
		Spring	0.582	-	
Mission-Aransas	0.677***	Summer	**0.848**	0.652	-
		Fall	**0.763**	**0.845**	0.502
		Spring	0.565	-	
Nueces	0.647***	Summer	**0.802**	0.627	-
		Fall	**0.700**	**0.873**	0.450
		Spring	0.493	-	
Upper Laguna Madre	0.516***	Summer	**0.651**	0.420	-
		Fall	0.515	**0.659**	0.381
		Spring	0.382	-	
Lower Laguna Madre	0.532***	Summer	**0.609**	**0.572**	-
		Fall	**0.629**	**0.764**	0.302

Table 3. Analysis of Similarity results of community structure within each estuary across seasons. Global R by Estuary, *** = $p < 0.001$, pairwise comparison R values by season (significant pairwise R values > than the Global R in bold).

http://dx.doi.org/10.5772/52313

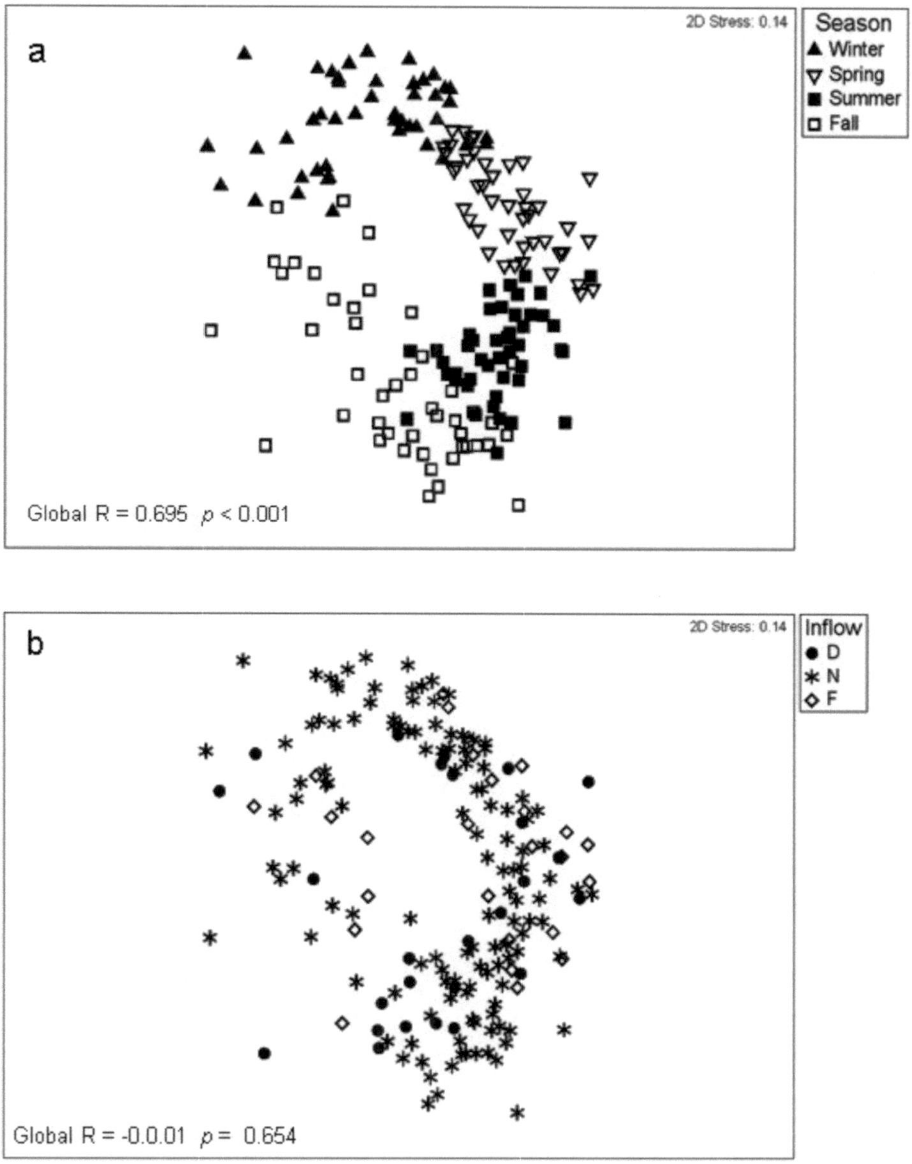

Figure 3. Multidimensional scaling (2D) configuration of bag seine community structure from a representative estuary (Lavaca-Colorado) overlaid with (a) Season, and (b) Inflow Condition. Season of collection defined as: Winter (Dec, Jan, Feb); Spring (Mar, Apr, May); Summer (Jun, Jul, Aug); and Fall (Sep, Oct, Nov). Inflow Condition designations; D = Drought, N = Normal, F = Flood. Global R values for each Analysis of Similarity test included.

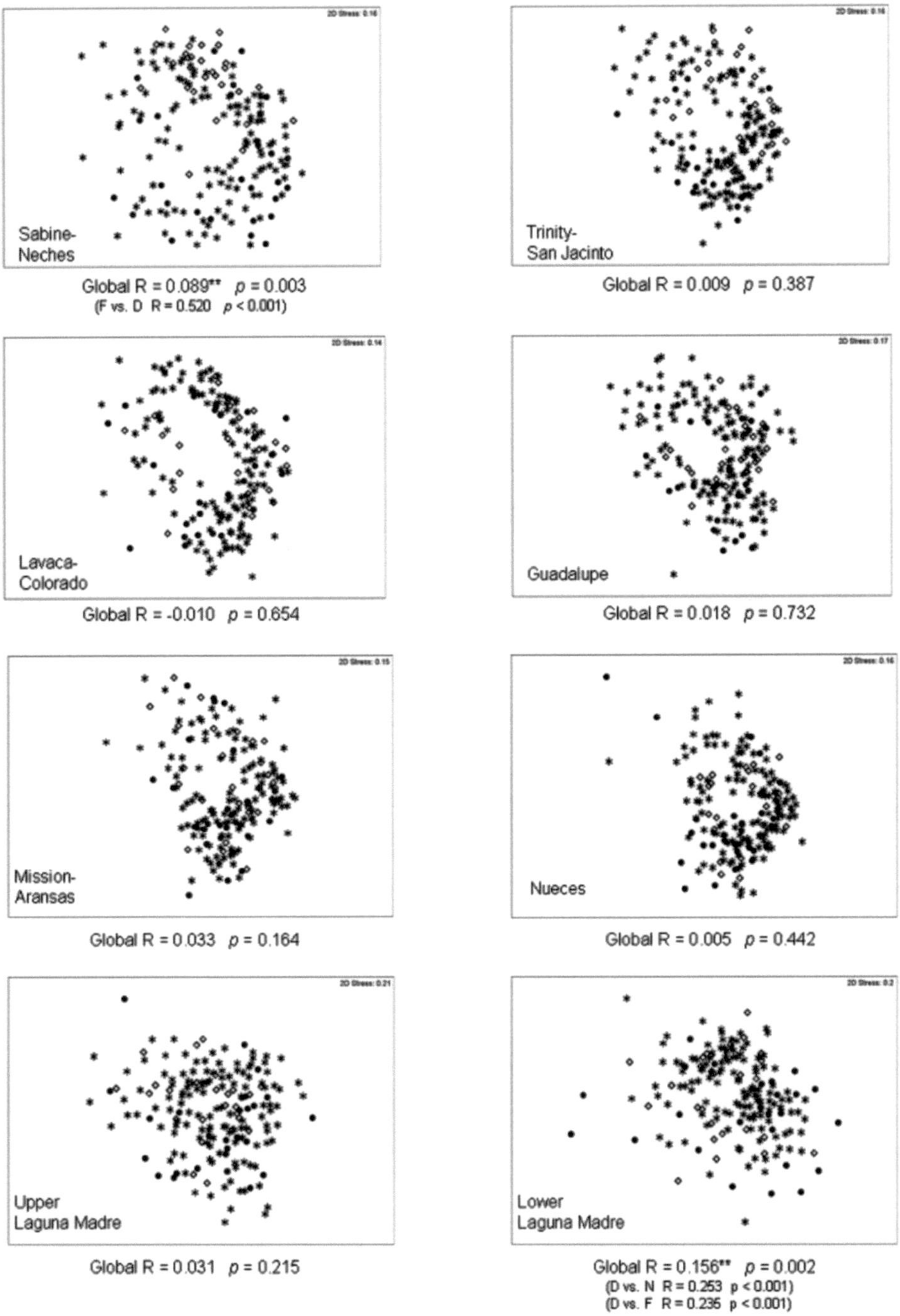

Figure 4. Multidimensional scaling (2D) configurations of bag seine derived community structure for each Texas estuary identifying the categorical Inflow Condition, with the Global R values for the seasonal Analysis of Similarity test. Inflow Condition symbols follow Figure 3.

While estuary-specific assemblages do not appear to be responding to synoptic inflow events, the inclusive role of salinity on overall community structure can be seen in Table 4. Across the estuaries, patterns of community structure roughly matched the NE-SW salinity gradient present on the Texas coast, with the freshest estuaries on the upper coast having significantly different communities than the more saline estuaries found on the lower coast. The middle coast estuaries (Lavaca-Colorado, Guadalupe, and Mission-Aransas) showed the greatest degree of overlap in their community structure. Of the hundreds of taxa that constitute the nekton recorded with the bag seines, differences among the estuaries could be explained by examining only a fraction of this total. Abundance levels of 36 taxa accounted for the vast majority of the individuals found in each estuary, ranging from a low of 99.5 % in the Lavaca-Colorado system to a high of 100% in the Nueces Estuary (Table 5). Of the commercially and recreationally important species currently used for TxEMP modeling, only blue crab, white shrimp, and brown shrimp contributed substantially to nekton community structure patterns. Other taxa used for TxEMP either contributed little to the overall community (e.g., red drum ranked no higher than 12th from any estuary; spotted seatrout no higher than 14th) or were identified as a characteristic component from only a single estuary (southern flounder contributed to at least 1% of the community structure only in the Sabine-Neches estuary). Black drum were not identified as a significant component from any estuary. Community structure across the salinity gradient of estuaries appears to be driven by the relative proportion of only a few oligohaline (e.g., Atlantic croaker, bay anchovy *Anchoa mitchilli*, and Gulf menhaden *Brevoortia patronus*) and polyhaline to metahaline taxa (pinfish, Gulf killifish *Fundulus grandis*, sheepshead minnow, longnose killifish *F. similis*, and grass shrimp *Palaemonetes* spp.). Ubiquitous euryhaline taxa that were identified at equivalent ranks across the coastal salinity gradient included blue crab, striped mullet *Mugil cephalus*, spot *Leiostomus xanthurus*, brown shrimp, white shrimp, and silversides *Menidia* spp.

Estuary	Sabine-Neches	Trinity-San Jacinto	Lavaca-Colorado	Guadalupe	Mission-Aransas	Nueces	Upper Laguna Madre
Trinity-San Jacinto	0.299	-					
Lavaca-Colorado	0.324	0.081	-				
Guadalupe	**0.686**	0.423	0.347	-			
Mission-Aransas	**0.555**	0.283	0.213	0.295	-		
Nueces	**0.710**	0.371	0.335	0.418	0.116	-	
Upper Laguna Madre	**0.875**	**0.806**	**0.727**	0.403	**0.579**	**0.679**	-
Lower Laguna Madre	**0.784**	**0.578**	**0.527**	0.324	0.381	0.336	**0.485**

Table 4. Matrix of pairwise comparison R values for the Analysis of Similarity test of community structure among estuaries. Global R = 0.455, $p < 0.001$; significant pairwise R values > than the Global R in bold.

Species	Sabine-Neches	Trinity-San Jacinto	Lavaca-Colorado	Guadalupe	Mission-Aransas	Nueces	Upper Laguna Madre	Lower Laguna Madre
Micropogonias undulatus	1	2	4	12	13	17		14
*Callinectes sapidus**	2	1	2	5	3	2	7	5
Anchoa mitchilli	3	5	10	11	11	13	8	17
Brevoortia patronus	4	3	11	16				
Menidia beryllina/ peninsulae	5	4	3	2	2	5	2	8
Mugil cephalus	6	6		8	10	10	10	9
*Litopenaeus setiferus**	7	10	6	15	12	14		13
Palaemonetes spp.	8	7	1	1	1	6	3	12
Leiostomus xanthurus	9	8	9	9	8	7	11	6
*Farfantepenaeus aztecus**	10	9	5	6	7	8	9	3
Lagodon rhomboides	11	12	8	3	4	3	6	1
*Sciaenops ocellatus**	12	16	15	14	14	16	16	18
*Paralichthys lethostigma**	13							
Cynoscion arenarius	14	17	18					
Fundulus similis		13	7	7	6	4	4	4
Fundulus grandis		14	13	10	9	9	5	15
*Cynoscion nebulosus**				17	17	18	14	
Gobiosoma bosc				18		19		
Eucinostomus argenteus				19	18			16
Cyprinodon variegatus		11	12	4	5	1	1	2
Mugil curema		15	14	13	15	15	15	11
Citharichthys spilopterus		18	19					

Species	Sabine-Neches	Trinity-San Jacinto	Lavaca-Colorado	Guadalupe	Mission-Aransas	Nueces	Upper Laguna Madre	Lower Laguna Madre
Callinectes similis		19				12		10
Menticirrhus americanus		20	16					
Arius felis		21	17					
Lucania parva							12	
Farfantepenaeus duorarum					16	11		7
Syngnathus scovelli							13	
Percent Total Abundance	99.7	99.7	99.6	99.5	99.8	100	99.8	99.7

Table 5. Rank order of the nekton taxa contributing to the top 90% of average similarity within each estuarine assemblage. Species identified by an asterisk (*) represent commercially or recreationally important target species currently used in TxEMP modeling. Blank entries represent taxa recorded from each estuary, but their overall contribution to the community in that estuary is less than 1%.

4. Discussion

Long-term data sets are fundamental to an understanding of factors that regulate system level processes, because the inherent complexity and variability of open natural systems make it difficult to establish causal relationships between and among the important components. These data are needed to ensure that the environmental conditions which potentially can lead to dramatic fluctuations in observed nekton abundance levels are recorded at least once and preferably several times [24]. Decadal-scale continuous records of biological data utilizing uniform sampling strategies are the exception rather than the rule for most estuarine and coastal realms [25]. Many of the estuarine studies that do take into account the spatial and temporal aspects of the physical environment often utilize commercial catch and effort records [26-30], and these catch per unit effort (CPUE) indices of abundance are not without problems. Technological advances and external economic factors which can directly affect actual effort are either poorly documented, or often entirely dismissed. Circumventing some of the inherent problems associated with fishery-dependent CPUE indices, TPWD utilizes fisheries-independent sampling methodologies to assemble long-term data sets of estuarine biotic and abiotic structure [21]. Besides providing resource managers with uniform information that is reliably documented and collected under standardized sampling designs and techniques, these long-term data sets offer the antithesis to short-term management solutions dictated by monetary constraints that emphasize research and monitoring projects of limited temporal and spatial duration [31].

The current TxEMP methodology which uses salinity as a proxy for FWI to establish inflow–species spatial relationships has demonstrated varying levels of correspondence between abundance and salinity gradients [14, 32-33]. In estuaries receiving substantial inflows that facilitate defined salinity zonations (e.g., Trinity-San Jacinto and Guadalupe), peak densities of many target species were spatially correlated with specific salinity zones. Conversely, in estuaries receiving lower amounts of inflows (e.g., Nueces and Laguna Madre), well defined salinity zonations were either dramatically compressed into the upper-most reaches of the estuary, or absent altogether, and consequently these same target species were far less associated with their recognized salinity preferences. Expanding the spatial scale beyond individual estuaries and using the bag seine information to encompass the entire nekton community, the present analysis shows that a much lower degree of correspondence exists between the synoptic-scale FWI signal and community assemblage. While this general lack of correspondence between the motile nekton and FWI may seem contradictory to the reported positive flow effects on fisheries abundance [29-30, 34-35], similar neutral responses by fisheries to FWI have been reported in other studies conducted at equivalent spatial scales as was used for this study. For example, in East Bay, Florida, Livingston et al. [2] found that river flow and primary production were associated mainly with changes in the communities at the lower trophic levels (herbivores and omnivore), whereas the carnivores (e.g., spotted seatrout, southern flounder, and red drum) were associated primarily with other animal trophic interactions. Their study showed that salinity changes were only indirectly involved in biological interactions at the highest trophic levels. Similarly, Griffiths [36] showed that yellowfin bream *Acanthopagrus australis* (the functional equivalent of pinfish used for this study) and striped mullet were generally resilient to salinity perturbations in Shellharbour Lagoon, Australia. Neutral responses to fluctuating salinities is not exclusive to finfish, as both Kimmerer [37] reporting on California bay shrimp *Crangon franciscorum* (functional equivalent of white shrimp) in the northern San Francisco Estuary, California, and Rozas et al. [11] working with brown shrimp in Breton Sound, Louisiana, each showed a general de-coupling of abundance levels and hydrologic conditions. Similarly, increases in temporal scales have also revealed a general de-coupling between abundance levels and FWI, as both spot and Atlantic croaker did not correlate with year-to-year variation of river discharge in Apalachicola Bay, Florida [38]. Weinstein et al. [39] also showed that shallow-water fish assemblages in the Cape Fear River estuary were not affected by annual differences in river discharge.

The repeatability of species assemblage composition and abundances from year to year across the salinity spectrum in the estuarine systems along the Texas coast is one of the most prominent features of this study. From Figure 4, it is clear that an orderly seasonal succession in abundance and species composition of the dominant components confirms the many published accounts of annually repeating community structure from a variety of locations [40-43]. A common theme found within these studies is that the identification of quite specific arrival times, or dates of first occurrence within each season of recruitment, can be shown for a number of taxa, regardless of the hydrologic conditions within an estuary at the time of recruitment. Interannual variations in these dates of first occurrence are typically small, suggesting that temporal stability of assemblage structure may be more closely related to temperature [40] or seasonal photoperiods than to salinity. The current analysis shows that the greatest disparity in community composition, regardless of any underlying salinity level difference, involved comparisons across opposite seasons. These seasonal differences were steadfastly replicated year after year, in spite of the dramatically different levels of freshwater inflows producing temporally unpredictable flood and drought conditions. During these environmental extremes, no wholesale changes in community composition were noted; only changes in the relative abundances within a set of common taxa.

Absent from the current analysis is a recognition of the role of physical habitat in structuring nekton community compositions (reviewed in [44]). From Table 1, it is quite clear that major differences in the areal extent of fringing wetlands and submerged aquatic vegetation exists among the eight estuaries under investigation, and despite these obvious differences, the major nekton components of each community assemblage are, for the most part, the same limited suite of taxa (Table 5). Many estuarine organisms have increased (sometimes dramatically) abundances in areas closest to the freshwater source, and these same oligo- and mesohaline areas are noted for supporting much of the wetland habitats cited in Table 1. Even though the direct effect that FWI has on wetlands and the species that use them has not been definitely demonstrated [12], there is very good evidence that these relationships exist, at least for some size-groups or life history stages [7, 18, 36, 45]. Transient groups of young-of-the-year clupeiforms (Gulf menhaden, bay anchovy), perciforms (Atlantic croaker, spot, red drum, pinfish, spotted seatrout, and both species of mullets) and pleuronectiforms (southern flounder, bay whiff) that are spawned in deeper estuarine, nearshore, or offshore waters have all been shown to enter the shallow portions of estuaries and occur in very high densities in their recruitment and residence periods [17-19, 34,47]. Those nekton communities from the fresher, upper coast estuaries with large amounts of surrounding wetlands supported greater proportions of Atlantic croaker, bay anchovy, Gulf menhaden, and striped mullet than did the more southern estuaries with less fringing marsh. In the more saline estuaries, where seagrasses generally replace fringing marsh systems as one of the dominant structured habitats, the nekton communities identified by the bag seines were characterized by increases in cyprinidontiforms (sheepshead minnows and longnose killifish), grass shrimp, pinfish. Except for pinfish, all these taxa are estuarine-residents that do not recruit from nearshore or offshore spawning grounds, therefore the intermediate linkage between freshwater inflows and physical habitats may not be as important for these populations to be successful.

Many investigations have suggested that variability in estuarine production can be attributed either directly or indirectly to the fertilizing effects of freshwater input [1, 10, 48-49]. This estuarine 'agricultural model' is based on a mechanistic link between nutrient loading and increased phytoplankton production, ultimately leading to increased fisheries yields. Although Sutcliffe's [48] arguments have been disputed on interpretation and statistical grounds [46, 50], the concept persists and is fundamental to the implementation of the TxEMP methodology. Relating the flow effects to animal populations requires trophic transfer up the food web, and numerous studies have focused on the relative importance of 'top-down' vs. 'bottom-up' control of aquatic food webs [51-54]. A distinct dichotomy in the response of FWI controlling factors within estuarine systems appears to be that the herbivores and omnivores are more directly linked to physical and chemical controls associated implicitly with primary production ('bottom-up' regulation), whereas the carnivores (primary, secondary, and tertiary) are more closely associated with 'top-down' biological factors [2, 38]. Examination of the target taxa used for the TxEMP inflow-species relationships (Table 5) shows that all the vertebrates fall into the tertiary carnivore class, and the epibenthic macroinvertebrates are classified by as primary and secondary carnivores [2]. Omnivorous taxa that are more likely to have a more direct trophic linkage to the effects of FWI included pinfish, spot, striped mullet, white mullet, hardhead catfish *Arius felis*, and Gulf pipefish *Syngnathus scovelli*. Spot, a bottom-feeding perciform characteristic of the assemblage structure in every estuary, are potentially an important linkage between inflows and production because of their ability to regulate benthic invertebrates [55]. The connection between the benthos and FWI associated production appears show a much stronger mechanistic link [38, 56-57], although the benthic environment is represented only by eastern oyster within the current modeling paradigm. Thus, evaluating the biological effects of FWI within Texas estuaries is currently dependent upon taxa that empirically have been shown to display the weakest mechanistic couplings.

The time steps involved in TxEMP FWI modeling are on the order of a calendar month, whereas the nekton appear to be operating on the order of months (the seasonal signal was clearly evident in every estuary across the coast) to a calendar year (the repeating pattern of seasonal compositions common to each estuary resulted in the circular configuration of the samples seen in Figure 3a). Conversely, the drought and flood FWI signal common to the entire Texas coast is due by climate-level drivers operating at fundamental frequencies of approximately 11, 5, and 3.5 years [58]. While all of the commercially and recreationally important finfish used to quantify optimal FWI are long lived species and can contribute a number of different year-classes to the nekton community, the macroinvertebrates used by TxEMP all have life spans less than even the shortest frequency inflow signal driver. The shrimp species are all considered annual species [15], with maximum life spans from 18 months to 2 years, whereas blue crabs are reported to have a life span approaching 3 years. For the shrimp species abundant throughout the Texas coast, physical timing of their recruitment periods appear to be more in synchrony with species-specific temperature ranges instead of estuarine salinity requirements. Brown shrimp recruit to the estuary from February through May, while white shrimp typically show up from June through October. Once recruited from offshore spawning areas, white shrimp juveniles can migrate farther into the less saline waters of the upper estuary because they are more tolerant of lower salinities than the other shrimp species. This pattern is evidenced by the higher rank abundance values for white shrimp seen in the less saline upper coast estuaries, whereas rank values for brown shrimp were higher in the more saline lower coast estuaries. While the interaction of available habitat and salinity tolerance levels can therefore aid in successful recruitment, the annual frequency in shrimp spawning does not appear to be closely tied to the multiyear to decadal frequencies of the inflows. These observations conform to the conclusions of Allen and Barker [19], in that 'responses of populations to major changes in the estuarine environment are more strongly expressed as alterations in the magnitude than in the timing of habitat utilization'.

Because estuarine fishes have evolved to exploit one of the more physiologically challenging environments, it should not too surprising that they do not appear to be dramatically responding to the synoptic-scale inflow events that are currently used to quantify 'ecological health'. Even when utilizing species ranks to adjust for any gross differences in relative abundance among estuaries (e.g., biomass), the present analysis reveals that in each estuary, the contributions of a very limited number of taxa were strikingly similar. To evaluate the functional role of freshwater inflow into estuaries and determine estuarine FWI needs for the future, incorporating more sensitive 'measuring stick' organisms are recommended. One way this could be accomplished is to incorporate a greater range of trophic structure, utilizing some of the lower trophic level taxa that constituted a majority portion of the community assemblage (e.g., Gulf menhaden, bay anchovy, Gulf killifish, striped mullet, sheepshead minnow), or base the FWI modeling on taxa that appear to show a definite salinity response (e.g., Atlantic croaker, longnose killifish, white mullet, pinfish). Still another more challenging option would be to move down the trophic food web and index measures of 'estuarine health' to benthic taxa that show more direct mechanistic linkages to FWI.

Acknowledgements

I am sincerely indebted to the more than twenty years of field staff and technicians at the Coastal Fisheries Division of Texas Parks and Wildlife Department for their diligent collection of the biotic and abiotic parameters used for this study. This project was never explicitly funded by research grants, but I gratefully acknowledge the continued support of Sportfish Restoration Funds, without which the time for data synthesis and interpretation would not have been possible.

Author details

James M. Tolan*

Address all correspondence to: james.tolan@tpwd.state.tx.us

Texas Parks and Wildlife Department, Coastal Fisheries Division, Natural Resource Center 2501, Unit 5846, Corpus Christi, TX, USA

References

[1] Flint, R.W. Long-term estuarine variability and associated biological response. Estuaries 1985;8 158-169.

[2] Livingston, R.J., N. Xufeng, F.G. Lewis, III, and G.C. Woodsum. Freshwater input to a Gulf estuary: long-term control of trophic organization. Ecological Applications 1997;7 277-299.

[3] Logeragan, N.R., and S.E. Bunn. River flows and estuarine ecosystems: implications for coastal fisheries from a review and case study of the Logan River, southeast Australia. Australian Journal of Ecology 1999;24 431-440.

[4] Alber, M. A conceptual model of estuarine freshwater inflow management. Estuaries 2002;25 1246-1261.

[5] Vörösmarty, C.J., and D. Sahagian. Anthropogenic disturbance of the terrestrial water cycle. BioScience 2000;50 753-765.

[6] Dynesius, M., and C. Nilsson. Fragmentation and flow regulation of river systems in the northern third of the world. Science 1994;266 753-762.

[7] Sklar, F.H., and J.A. Browder. Coastal environmental impacts brought about by alterations to freshwater flow in the Gulf of Mexico. Environmental Management 1998;22 547-562.

[8] Rijsberman, F.R. Water scarcity: Fact or fiction? Agricultural Water Management 2006;80 5-22.

[9] Skreslet, S. Freshwater outflow in relation to space and time dimensions of complex ecological interactions in coastal waters. In: S. Skreslet (ed.) The role of freshwater outflow in coastal marine ecosystems. Berlin, Germany: Springer-Verlag; 1986. p3-12.

[10] Mallin, M.A., H.W Paerl, J. Rudek, and P.W. Bates. Regulation of estuarine primary productivity by watershed rainfall and river flow. Marine Ecology Progress Series 1993;93 199-203.

[11] Rozas, L.P., T.J. Minello, I. Munuera-Fernandez, B. Fry, and B. Wissel. 2005. Macrofaunal distributions and habitat change following winter-spring releases of freshwater into the Breton Sound estuary, Louisiana (USA). Estuarine, Coastal and Shelf Science 65:319-336.

[12] Longley, W.L., editor. Freshwater inflows to Texas bays and estuaries: Ecological relationships and methods for determination of needs. Texas Water Development Board and Texas Parks and Wildlife Department, Austin, Texas. 1994.

[13] Powell, G.L., and J. Matsumoto. Texas estuarine mathematical programming model: a tool for freshwater inflow management. In: K.R. Dyer and R.J. Orth (eds.) Changes and fluxes in estuaries. Fredensborg, Denmark: Olsen and Olsen; 1994. p401-406.

[14] Pulich, W. Jr, J. Tolan, W.Y. Lee, and W. Alvis. Freshwater inflow recommendation for the Nueces Estuary. http://www.tpwd.state.tx.us/landwater/water/conservation/freshwater_inflow/nueces/ (accessed 20 Feb 2010).

[15] Patillo, M.E., T.E. Czapla, D.M. Nelson, and M.E. Monaco. Distribution and abundance of fishes and invertebrates in Gulf of Mexico estuaries, Volume II: Species life

history summaries. ELMR Report Number 11. NOAA/NOS Strategic Assessment Division, Silver Spring, MD. 1997.

[16] Able, K.W. A re-examination of fish estuarine dependence: Evidence for connectivity between estuarine and ocean habitats. Estuarine, Coastal and Shelf Science 2005;64 5-17.

[17] Weinstein, M.P., and H.A. Brooks. Comparative ecology of nekton residing in a tidal creek and adjacent seagrass meadow: community composition and structure. Marine Ecology Progress Series 1983;12 15-27.

[18] Rogers, S.G., T.E Targett, and S.B. Van Sant. Fish-nursery use in Georgia salt-marsh estuaries: the influence of springtime freshwater conditions. Transactions of the American Fisheries Society 1984;113 595-606.

[19] Allen, D.M., and D.L. Barker. Interannual variation in larval fish recruitment to estuarine epibenthic habitats. Marine Ecology Progress Series 1990;63 113-125.

[20] Hare, J.A., and K.W. Able. Mechanistic links between climate and fisheries along the east coast of the United States: explaining population outbursts of Atlantic croaker (Micropogonias undulatus). Fisheries Oceanography 2007;16 31-45.

[21] Martinez-Andrade, F., P. Campbell, and B. Fuls. Trends in relative abundance and size of selected finfishes and shellfishes along the Texas Coast: November 1975-December 2003. Management Data Series No. 232, Texas Parks and Wildlife Department, Coastal Fisheries Division. Austin, Texas. 2005.

[22] McKee, D.A. Fishes of the Laguna Madre: A guide for anglers and naturalists. College Station, Texas: Texas A&M University Press. 2008.

[23] Clarke, K.R., and R.M. Warwick. Change in marine communities: an approach to statistical analysis and interpretation, 2nd edition. Plymouth : PRIMER-E. 2001.

[24] Kruskal, J.B. Multidimensional scaling by optimizing goodness of fit to a non-metric hypothesis. Psychometrika 1964;29 1-27.

[25] Rose, K.A., and J.K. Summers. Relationships among long-term fisheries abundances, hydrographic variables, and gross pollution indicators in northeastern U.S. estuaries. Fisheries Oceanography 1992;1 281-293.

[26] Wolfe, D.A., M.A. Champ, D.A Flemer, and A.J. Mearns. Long-term biological data sets: their role in research, monitoring, and management of estuarine and coastal marine systems. Estuaries 1987;10 181-193.

[27] Summers, J.K., T.T. Polgar, J.A. Tarr, K.A. Rose, D.G. Heimbuch, J. McCurley, R.A. Cummins, G.F. Johnson, K.T. Yetman, and G.T. DiNardo. Reconstruction of long-term time series for commercial fisheries abundance and estuarine pollution loadings. Estuaries 1985;8 114-124.

[28] Pearson, T.H., and P.R.O. Barnett. Long-term changes in benthic populations in some west European coastal areas. Estuaries 1987;10 220-226.

[29] Houde, E.D., and E.S. Rutherford. Recent trends in estuaries fisheries-predictions of fish production and yield. Estuaries 1993;16 161-176.

[30] Jassby, A.D., W.J. Kimmerer, S.G. Monismith, C. Armor, J.E. Cloern, T.M. Powell, J.R. Schubel, and T.J. Vendlinski. Isohaline position as a habitat indicator for estuarine populations. Ecological Applications 1995;5 272-289.

[31] Diop, H., W.R. Keithly, Jr., R.F. Kazmierczak, Jr., and R.F. Shaw. Predicting the abundance of white shrimp (Litopenaeus setiferus) from environmental parameters and previous life stage. Fisheries Research 2007;86 31-41.

[32] Champ, M.A. Monitoring: Painting a moving train. Sea Technology 1986;27 73.

[33] Pulich, W. Jr, J. Tolan, W.Y. Lee, and W. Alvis. Freshwater inflow recommendation for the Nueces Estuary. http://www.tpwd.state.tx.us/landwater/water/conservation/freshwater_inflow/nueces/ (accessed 20 Feb 2010).

[34] Tolan, J.M., W.Y. Lee, G. Chen, and D. Buzan. Freshwater inflow recommendation for the Laguna Madre Estuary system. Texas Parks and Wildlife Department. Austin, Texas. 2004.

[35] Copeland, B.J. Effects of decreased river flow on estuarine ecology. Journal of the Water Pollution Control Federation 1966;38 1831-1839.

[36] Weinstein, M.P., and M.P. Walters. Growth, survival and production in young-of-year populations of Leiostomous xanthurus Lacépéde residing in tidal creeks. Estuaries 1981;4 185-197.

[37] Griffiths, S.P. Factors influencing fish composition in an Australian intermittently open estuary. Is stability salinity-dependent? Estuarine, Coastal and Shelf Science 2002;52 739-751.

[38] Kimmerer, W.J. Effects of freshwater flow on abundance of estuarine organisms: physical effects or trophic linkages? Marine Ecology Progress Series 2002;243 39-55.

[39] Kobylinski, G.J., and P.F. Sheridan. Distribution, abundance, feeding and long-term fluctuations of spot, Leiostomus xanthurus, and croaker, Micropogonias undulates, in Apalachicola Bay, Florida, 1972-1977. Contributions in Marine Science 1971;22 149-161.

[40] Weinstein, M.P., S.L. Weiss, and W.F. Walters. Multiple determinants of community structure in shallow marsh habitats, Cape Fear River estuary, North Carolina, USA. Marine Biology 1980;58 227-243

[41] McGovern, J.C., and C.A. Wenner. Seasonal recruitment of larval and juvenile fishes into impounded and non-impounded marshes. Wetlands 1990;10 203-221.

[42] Witting, D.A., K.W. Able, and M.P. Fahay. Larval fishes of a Middle Atlantic Bight estuary: assemblage structure and temporal stability. Canadian Journal of Fisheries and Aquatic Sciences 1999;56 222-230.

[43] Hagan, S.M., and K.W. Able. Seasonal changes of the pelagic fish assemblage in a temperate estuary. Estuarine, Coastal and Shelf Science 2003;56 15-29.

[44] Tolan, J.M. Larval fish assemblage response to freshwater inflow: a synthesis of five years of ichthyoplankton monitoring within Nueces Bay, Texas. Bulletin of Marine Science 2008;82 275-296.

[45] Minello, T.J. Nekton densities in shallow estuarine habitats of Texas and Louisiana and the identification of essential fish habitat. In L. Benaka (ed.) Fish habitat: Essential fish habitat and habitat restoration. American Fisheries Society, Symposium 22. Bethesda, MD. 1999. p43-75.

[46] Sinclair, M., G.L. Bugden, C.L. Tang, J.C. Therriault, and P.A. Yeats. Assessment of effects of freshwater runoff variability on fisheries production in coastal waters. In S. Skreslet (ed.) The role of freshwater outflow in coastal marine ecosystems. Berlin, Germany: Springer-Verlag. 1986. p139-160.

[47] Martino, E.J., and K.W. Able. Fish assemblage across the marine to low salinity transition zone of a temperate estuary. Estuarine, Coastal and Shelf Science 2003;56 969-987.

[48] Sutcliffe, W.H., Jr. Some relation of land drainage, particulate matter, and fish catch in to eastern Canadian bays. Journal of the Fisheries Research Board of Canada 1972;29 357-362.

[49] Cloern, J.E., A.E. Alpine, B.E. Cole, R.L.J Wong, J.F. Arthur, and M.D. Ball. 1983. River discharge controls phytoplankton dynamics in the northern San Francisco Bay estuary. Estuarine, Coastal and Shelf Science 1983;16 415-429.

[50] Drinkwater, K.F., and R.A. Myers. Testing predictions of marine fish and shellfish landings from environmental variables. Canadian Journal of Fisheries and Aquatic Sciences 1987;44 1568-1573.

[51] McQueen, D.J., R.S. Johannes, J.R. Post, T.J. Stewart, and D.R.S. Lean. Bottom-up and top-down impacts on freshwater pelagic community structure. Ecological Monographs 1989;59 289-309.

[52] Menge, B.A. Community regulation: under what conditions are bottom-up factors important on rocky shores. Ecology 1992;73 755-765.

[53] Flinkman, J., E. Aro, I. Vuorinen, and M. Viitasalo. Changes in northern Baltic zooplankton and herring nutrition from 1980s to 1990s: top-down and bottom-up processes at work. Marine Ecology Progress Series 1998;165 127-136.

[54] Micheli, F. Eutrophication, fisheries, and consumer-resource dynamics in marine pelagic ecosystems. Science 1999;285 1396-1398.

[55] Killam, K.A., R.J. Hochberg, and E.C. Rzemiem. Synthesis of basic life histories of Tampa Bay species. Tampa Bay National Estuary Program, Technical Publication Number 10-92. 1992.

[56] Nilsson, P., B. Jönsson, I.L. Swanberg, and K. Sundbäck. Response of a marine shallow-water sediment system to an increased load of inorganic nutrients. Marine Ecology Progress Series 1991;71 275-290.

[57] Montagna, P.A., and R.D. Kalke. The effect of freshwater inflow on meiofaunal and macrofaunal populations in the Guadalupe and Nueces Estuaries. Estuaries 1992;15 307-326.

[58] Tolan, J.M. El Nino-Southern Oscillation impacts translated to the watershed scale: salinity patterns along the Texas Gulf Coast, 1982 to 2004. Estuarine, Coastal and Shelf Science 2007;72 247-260.

Chapter 5

Coastal Reservoir Strategy and Its Applications

Shuqing Yang, Jianli Liu, Pengzhi Lin and Changbo Jiang

Additional information is available at the end of the chapter

http://dx.doi.org/10.5772/52315

1. Introduction

While the world's population tripled in the 20th century, the use of renewable water resources has grown six-fold [1]. It is estimated that the world population will enlarge by another 40 to 50 % in the following fifty years. The demand for water will be increasing resulted by the population growth combined with industrialization and urbanization, which will have serious consequences on the environment. According to WHO/UNICEF Joint Monitoring Programme (JMP) (2012 Update), 780 million people lack access to an improved water source; approximately one in nine people [2]. Water stress causes deterioration of fresh water resources in terms of quantity (aquifer over-exploitation, dry rivers, etc.) and quality (eutrophication, organic matter pollution, saline intrusion, etc.). In the Developing World, women and children walk miles to get water. The UN estimates that the average is 40 pounds of water carried 4 miles (18 kg-6 km). This takes hours, people can't attend school/work, deforms the spine and can leave women vulnerable to assault [3]. Figure 1 showed the state of water shortages based on synthetic evaluation of water management using for agriculture in 2007.

As about 66% of the world population would be confronted with water-shortage by 2025 [5]; the water from aquifers, which provide water for one-third of the world's population, are being used out before nature can complement them [6]. Water scarcity is already a focus of attention all over the world [7]. For example, the Southwest and Midwest areas of the USA and Australia are vulnerable to water scarcity [8]. In Australia, water allocations for irrigation have caused a conspicuous decline in rainfall and runoff in the past decades [9]. Figure 2 shows drinking water as the substance to transform in a social life. Many indigenous peoples also depend very much on natural resources and live in ecosystems especially vulnerable to the effects of climate change, such as small island developing states, arctic regions and high altitudes [10].

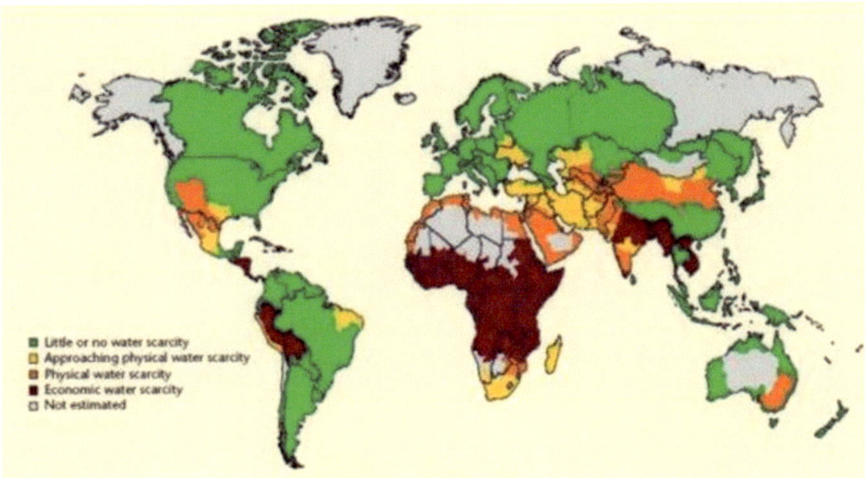

Figure 1. Water scarcity based on synthetic evaluation of water management using for Agriculture in 2007 [4]

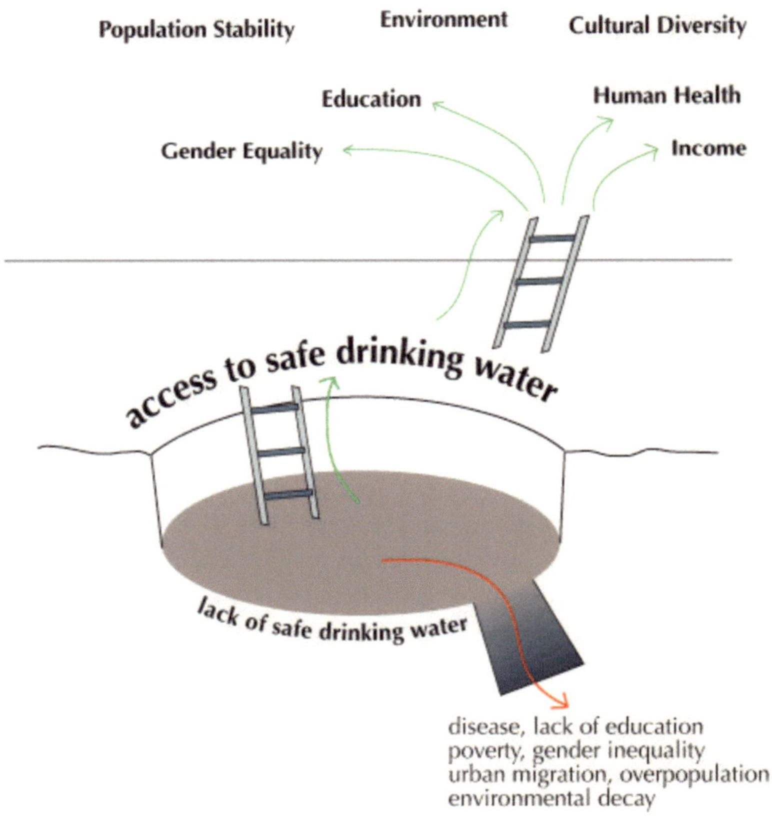

Figure 2. Drinking water as the substance to transform a life [11]

The shortage of fresh water reserves is a world-wide problem nowadays, which can lead to a global crisis. Firstly, it can cause agricultural crisis. 70–85 percent of water use is for agriculture, and it is assessed that the most imperilling agricultural growth is 20 percent of global grain production won't have enough water to irrigate in the future. Connected with spatial and temporal variations, water is not always available, which means there is scarcer water particularly for large concentrations of people consumption, enterprises and some other uses. Furthermore, it leads to environmental crisis. In Water and Nature, the more critical statement has been come out that the increasing of using water resource can reduce the amount of water available for industrial and agricultural development, more seriously that can exert a far-reaching influence on aquatic ecosystems and their dependent species. Environmental balances are disturbed and cannot play their roles like before [12]. More than 260 river basins are shared by two or more countries in the world. The lack of water can lead to transboundary tensions. As time passes, a series of conflicts come out, which increases regional instability, such as The Parana La Plata, the Aral Sea, the Jordan and the Danube.

Today, approximately 3 billion people — about half of the world's population — live within 200 kilometers of a coastline. By 2025,that figure is likely to double. In many countries, populations in coastal areas are growing faster than those in non-coastal areas. Take the example of China, the world's most populous nation. Of China's 1.2 billion people, close to 60% live in 12 coastal provinces, along the Yangtze River valley, and in two coastal municipalities — Shanghai and Tianjin. Along China's 18,000 kilometers of continental coastline, population densities average between 110 and 1,600 per square kilometer. In some coastal cities such as Shanghai, China's largest with 17 million inhabitants, population densities average over 2,000 per square kilometer. From 1990 to 2015, people flock to coastal cities continuously, especially in the past ten years, the number of migrated population is more and more bigger. Figure 3 [13] shows the population changes by 5 years (from 1990-2015) in Low elevation coastal zones (LECZ is a continuous strip along coastal areas whose altitude is less than 10m). In the United States, over 50 percent of the nation's population lives in only less than 20 percent of the U.S. coastal area (excluding Alaska) in 2010 [14]. Australia has experienced the same situation. Australia's coastal population has been growing faster than the population of the rest of the country for some time [15] (see Figure 4) and is expected to increase by another one million people over the next 15 years [16]. Up to 50 percent of the population in northern Africa and Bangladesh lives in coastal areas; along the Nile Delta, the population density reaches 500 to 1,000 people per square kilometre [17]. The high concentration of people in coastal regions has produced many economic benefits. But the combined effects of booming population growth and economic and technological development have aggravated the need for water.

Ban Ki-moon warns: "A shortage of water resources could spell increased conflicts in the future. Population growth will make the problem worse. So will the climate change. As the global economy grows, so will its thirst. Many more conflicts lie just over the horizon." [4]

With the current state of affairs, correct measures are needed to be taken to avoid the crisis to be worsening. There is an increasing awareness that our freshwater resources are limited and need to be noticed quantity and quality. "The time for solutions" is put forward in The 6th World Water Forum. This study summarized the overall situation of water solutions in the past and proposed a new waters strategy-coastal reservoir.

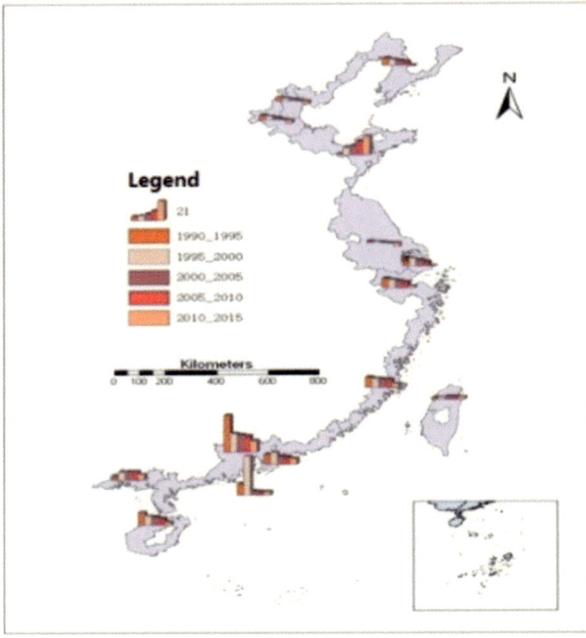

Figure 3. Coastal Population in China

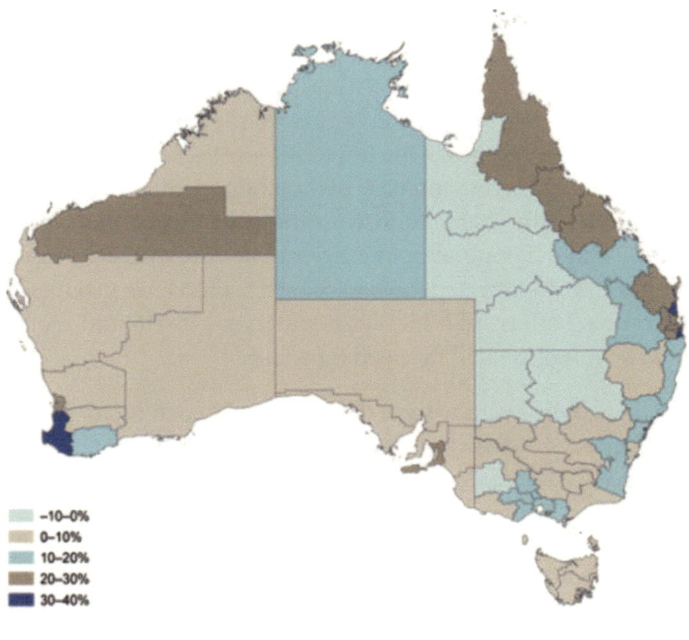

Figure 4. Coastal Population in Australia

2. Current water solutions

There are mainly several kinds of water supply options in the world, for example, groundwater, on-land reservoir, desalination of seawater, reuse of wastewater, Diversion of water from a remote source. They have their own characters.

Groundwater generally has higher chemical, physical and biological quality than surface waters. This higher quality means that groundwater usually requires lower levels of treatment, and the treatment process is therefore cheaper. But Groundwater is harder to locate than surface water. The location of groundwater may require the use of sophisticated equipment. But groundwater resources in the world are almost fully developed, and in some cases they are already considered overdeveloped. The groundwater sources that could potentially be developed in the world are so small that it would not be economically viable to do so. Therefore, the next preferable water supply source must be considered.

The water quality in on-land dams is higher than that of the water from all the alternative sources except groundwater. But surface waters are more susceptible to pollution than groundwater. They are difficult to protect from contamination (such as waterborne diseases and chemicals that enter from surface runoff and upstream discharges). The water quality is also not as consistent. For instance, turbidity and temperature fluctuate, often depending on the amount of precipitation. Surface water requires more treatment than groundwater and is therefore more expensive to utilise. This water supply source relies on rainfall. There are also usually a limited number of dam sites available because the construction of dams requires the correct combination of topography and geology. This is due to a lack of appropriate sites for the construction of dams. For example in Australia, after two centuries of development, there are almost no new dam sites left with good combination of topography and geology. Soil erosion also keeps reducing the storage capacity of the world's on-land reservoirs by more than 1% annually. For example, on-land reservoirs were subjected to very high siltation rates which are comparable to overseas extreme siltation rates in Australia [18]. In the future, soil erosion and reservoir sedimentation rates would be accelerated due to the severity of storms and rains as a result of global warming [10]. This means that almost all existing on-land reservoirs may lose its capacity in about 100 years. Therefore, the next water source option should be considered.

The first desalination plant by the multi-stage flash (MSF) evaporation process installed in 1964 in Lanzarote(Canary Islands) to over 700 desalination plants existing nowadays [19].Desalination technologies account for a worldwide production capacity of 24.5 million m^3/day [20], the cost decreased to \$0.50-0.80/m^3 desalinated water and even to 0.20-0.35 \$/m^3 for treatment of brackish water [21]. However, they have potential negative impacts [22]. These are mainly attributed to the concentrate and chemical discharges, which may impair coastal water quality and affect marine life, and air pollutant emissions attributed to the energy demand of the processes. A key concern of desalination plants are the concentrate and chemical discharges to the marine environment, which may have adverse effects on water and sediment quality, impair marine life and the functioning and intactness of coastal ecosystems [22]. WHO [23] provided a general overview on the composition and effects of the waste discharges in their guidance document. Lattemann [22, 24] and MEDRC [25] discussed details of the negative impacts. Reverse osmosis (RO), a commonly used desalination technology, is significantly more expensive than the standard treatment of fresh- water for potable use [20]. Furthermore, desalination technologies have the risk of oil pollution [26], Most of the noise is produced by the high-pressure pumps and by the turbines used for energy restoration [27].Desalination of seawater does not rely on rainfall (it is a climate resilient water source). Desalination plants are located on the coast, and as mentioned earlier this is where the majority of the water demand is expected to be located. But the treatment process would have high energy requirements as well as high operating and maintenance costs because of the high concentrations of contaminants that would have to be removed. For instance, seawater is usually 30 times more saline than the sewage effluent that would need to be treated to produce PRW (QWC, 2008). The concentrate stream produced by the desalination process can have detrimental environmental impacts. This could be a significant problem if the desalination plants were located on the coast, because the discharge would be released into a place, which is maybe a very sensitive marine environment. Therefore, other water supply options need to be considered.

Diversion of water from a remote source does not depend on local rainfall or local water quality. But this water source would have very high construction costs. For instance, the government of Queensland in Australia has considered diverting water supplies from North Eastern NSW, but this was found to be not economically viable. There were also numerous social, environmental and interstate issues that were considered to be insurmountable (QWC, 2008). Water could potentially be diverted from Northern Queensland. However, this would involve very high construction costs and should only be considered if no alternative water sources are available locally.

Reuse of wastewater does not depend directly on rainfall. But the levels of contaminants are higher than for the previous water supply options. Therefore, the cost of operating the treatment process would be higher. It is a common practice to discharge untreated sewage directly into bodies of water or put onto agricultural land, causing significant health and economic risks [28]. There could also be public perception problems with the reuse of wastewater (this is often referred to as the 'Yuk Factor').

Stormwater harvesting is used in house yards, streets and parks, but there are several shortages. CM Goonrey et al. exam the technical feasibility of using stormwater as an alternative supply source in an existing urban area [29]. Firstly, its storage capacity is very small due to the structural constraints; then its water quality may be not very good as the water comes from densely populated areas, and its cost is also very high as precious land is used to store the storm water.

Now in the world, many people try to find ways to solve water scarcity, however, on the other hand,too much stormwater and flood can't be neglected. The former management policies lead to vast amount of floodwater being discharged into the sea. Drought and lack of water emerge during dry season. It is obvious that new waters storages are in demand in the country. The coastal reservoir is to solve water storage of the next generation. It is a dam in the sea at the mouth of river or near it. It catches water in the bottom of the stream [30]. The building imitates the natural freshwater lake with a sustainable flow of fresh water from the river. Such reservoirs were already used in the Asian countries – Singapore, Hong Kong, China – and they are considered as very successful alternative to traditional dams.

"There is a water crisis today. But the crisis is not about having too little water to satisfy our needs. It is a crisis of managing water so badly that billions of people - and the environment - suffer badly." ——World Water Vision Report

---

3. Coastal reservoir

A Coastal Reservoir is a freshwater reservoir located in the sea at the mouth of a river with a sustainable annual river flow [31]. All a costal reservoir needs to be effective is an impermeable barrier between the fresh river water and the salty sea water. Yang outlined three guidelines for the successful construction of a coastal reservoir. The first guideline is separation, meaning the successful separation of clean river from polluted water and salt water. Next is protection, meaning the protection of collected fresh water against polluted river water and external pollution. Last is prevention, meaning the successfully prevention of salt water intrusion into the stored fresh water; weather that be through permeability or large tidal events.

Compared with water from seawater desalination processes, catchment runoff is a natural resource, which is cost-saving and water quality as the basic guarantee, is more closed to drinking water. Different from the on-land catchment runoff proposal, coastal reservoir harvests the catchment runoff in the sea, its water sources from a river and the reservoir has the potential to catch all runoff from a catchment. The coastal reservoir can be classified into various categories, in terms of location, barrage, and water quality etc. Existing freshwater lakes or lagoons on the shore can be regarded as special or natural coastal reservoirs. The main differences between coastal reservoirs and on-land reservoirs are summarized in Table1.

Item	On-land Reservoir	Coastal Reservoir
Dam-site	Valley(limited area)	Coast(inside/outside river mouth)
Water level	Above sea level	At sea level
Pressure	High pressure	Low pressure but with wave surge
Seepage	By head difference	By density difference
Pollutant	Land based	Land-based and seawater
Land acquisition	High	Low
Water supply	By gravity	By pump

Table 1. Differences between on-land Reservoirs and Coastal Reservoirs

Table 1 shows when constructed they share many similarities to inland reservoirs but due to the presence of water on either side which have almost the same density the hard barrier does not need to be constructed as strong as an inland reservoir does (material costs are a lot lower).

3.1. Existing coastal reservoirs

The first coastal reservoir was built in Zuider Zee, Netherlands in 1932, named Ijsselmeer with water area of 1240 km^2. At that time, people mainly enclose sea area to set up land with getting water as a fringe benefit. During the late 20th century and early 21st century, these are predominantly man made instances of coastal reservoirs with accelerated dilution of salt water to form viable storage and catchment of potable water.

The construction of these coastal reservoirs involves forming a dam wall, usually of solid material across the point where a river or lake enters the ocean. Salt water dammed in the reservoir is then pumped out to aid the speed in which water is diluted to form fresh water. As this reservoir is formed at sea level, maximum flow will occur and full use of the catchment inflows feeding the river or lake will be made. Ordinary storage reservoirs that involve river dams also house water from catchment inflows however these reservoirs are commonly located further upstream which allows all water below the dam to be released into the ocean. This type of Coastal Reservoir is being used in Singapore, South Korea, Hong Kong and China. Currently, there exist many coastal reservoirs in the world as listed in Table2.

Name	Catchment (km^2)	Dam length(m)	Capacity (million m^3)	Year completed	Country/river
Qingcaosha	66.26	48786	435	2011	China/ Yangtze
Saemanguem		33900	530	2010	South Korea
Sihwa	56.5	12400	323	1994	South Korea
Marina Barrage		350	42.5	2008	Singapore
Chenhang				1992	China/Shanghai
Yu Huan	166	1080	64.1	1998	China/Zhejiang
Baogang				1985	China/Shanghai
Plover Cove	45.9	2000	230	1968	Hong kong

Table 2. Existing Coastal Reservoirs in the World

3.2. The types and functions of coastal reservoir

The coastal reservoir can be classified into various categories, in terms of location, barrage, and water quality etc. According to the geographical location, it can be divided to estuary reservoir, intertidal reservoir, gulf reservoir. According to the water quality, it can be divided to drinking water reservoir with good quality, freshwater reservoir for agricultural/

industrial purpose with moderate quality, sewage reservoir, ballast water reservoir, etc. According to the dam body, it can be divided to concrete dam, earth dam and soft dam reservoirs. And it also can be divided to natural and aritificial, for example, Saemangeum coastal reservoir, South Korea, which is a huge project of the century.

Coastal reservoir can be used to provide water to three main parts, irrigation, industrial and domestic water using. For example, Chenhang Reservoir, which is located in Yangtze River Estuary of China, mainly provides drinking water for north area of Shanghai (see Figure.5). Baogang Reservoir, which is located in Yangtze River Estuary of China, plays an important role in providing industrial water to keep Baoshan Steel going on (see Figure5).

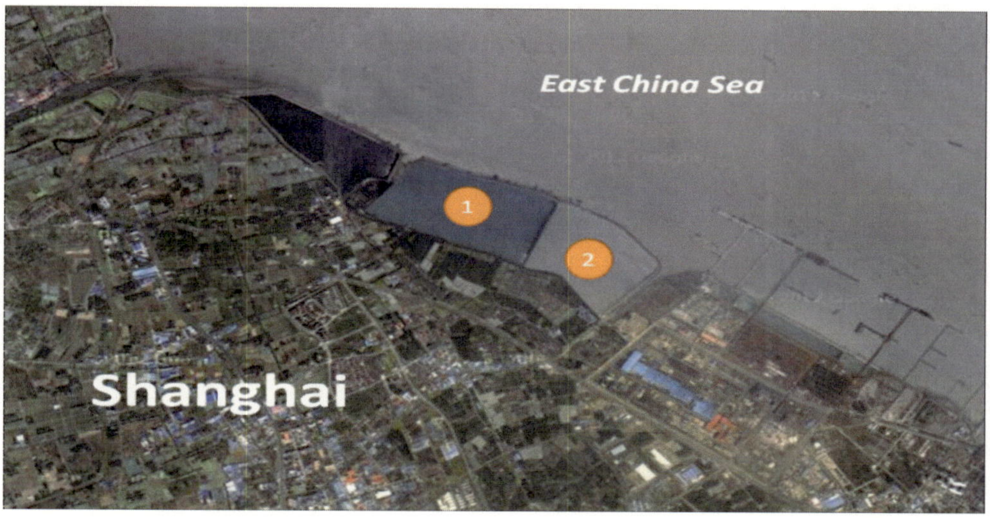

Figure 5. Baogang and Chenhang Reservoir (**1** is Baogang Reservoir and **2** is Chenhang Reservoir)

3.3. Plover cove reservoir in Hong Kong

Plover Cove Reservoir, located within Plover Cove Country Park, in the northeastern New Territories, is the largest reservoir in Hong Kong in terms of area, and the second-largest in terms of volume. It was the first in the world to construct a lake from an arm of the ocean. Its main dam was one of the largest in the world at the time of its construction, disconnecting Plover Cove from the sea.

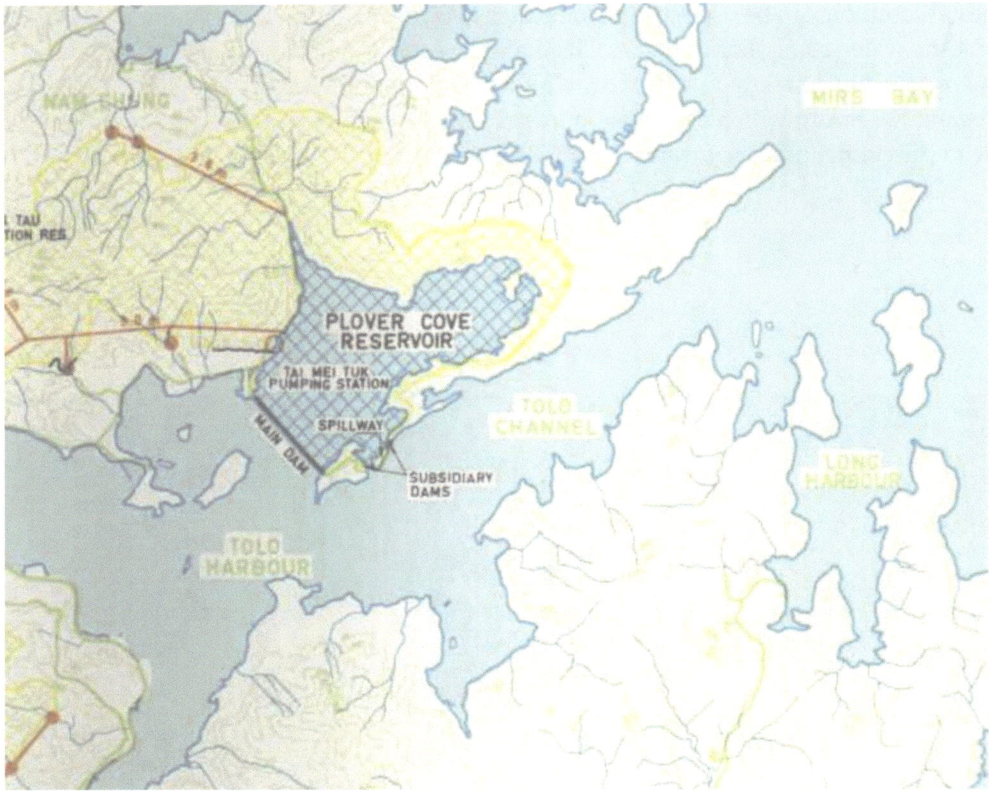

Figure 6. Plover Cove Reservoir in Hong Kong

The dam of the reservoir is 28m tall and approximately 2 km long, which was built by layers of sand and gravel. Besides rain from its catchment, it also stores water imported by pipes from Dongjiang. The Bride's Pool flows into the Plover Cove Reservoir. One main dam and three service dams were built to shut the cove off from the sea. The cove was then drained and was converted into a fresh-water lake.

The location was a former cove (bay, as the name suggests) and was a popular hiking site. Construction work commenced in 1960 and was completed in 1968, providing a capacity of 170 million m^3. Work on raising the height of the dams began in 1970. Upon completion in 1973, the reservoir capacity was increased to 230 million m^3 (see Figure 6).

http://dx.doi.org/10.5772/52315

3.4. Coastal reservoir in Singapore

Singapore is a tropical coastal city between the lines of longitude E103° 38′ and 104°05′, the latitude N 1°09′ and 1°29′ with population of 4 million and area of 680km^2. The mean annual rainfall is 2.4m. The total evapo-transpiration, infiltration, etc. loss is around 1.17m to 1.27m per year. At the end of 2000, total volume of water consumed in Singapore was 455.4 million m^3/year. Half of Singapore's water supply is imported from Malaysia across Johor strait through a Causeway, and the other half of water demand comes from its own reservoirs.

Due to its rapid economic and population growth in the past decades, the demand for potable water in Singapore increases steadily, the former way of water supply can't satisfy the need of people's living and production. To augment water supply, Singapore has built the Marina Barrage to develop further its rainfall (see Figure 7).

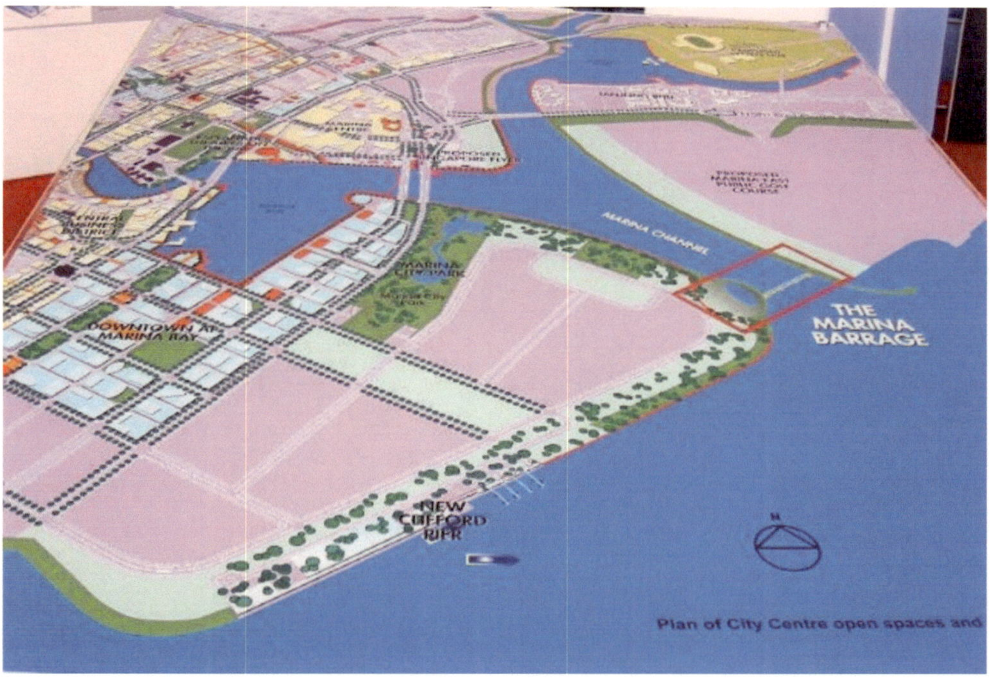

Figure 7. Marina Barrage in Singapore (Across Marina Channel at Marina South)

The Marina Barrage is a dam built across the 350-metre wide Marina Channel to keep out seawater. Marina Reservoir, together with Punggol and Serangoon reservoirs, increased Singapore's water catchment area from half to two-thirds of Singapore's land area in 2011. It creates a fresh water reservoir behind it at the same time acts as a tidal barrier that prevents high tides from causing flooding of inland low-lying areas. The Project is unique in that it is designed to achieve three aims: to act as a tidal barrier for flood control, to create a new reservoir to augment the water supply and to maintain a new body of freshwater at constant level in the heart of the city as a major lifestyle attraction. The Project has been carefully designed to blend in well with the environment, with guidance from Urban Redevelopment Authority (URA)'s Design Advisory Panel. Figure 8 shows the model of the Marina Barrage design, and Figure 9 shows the working principal of the Marina Barrage.

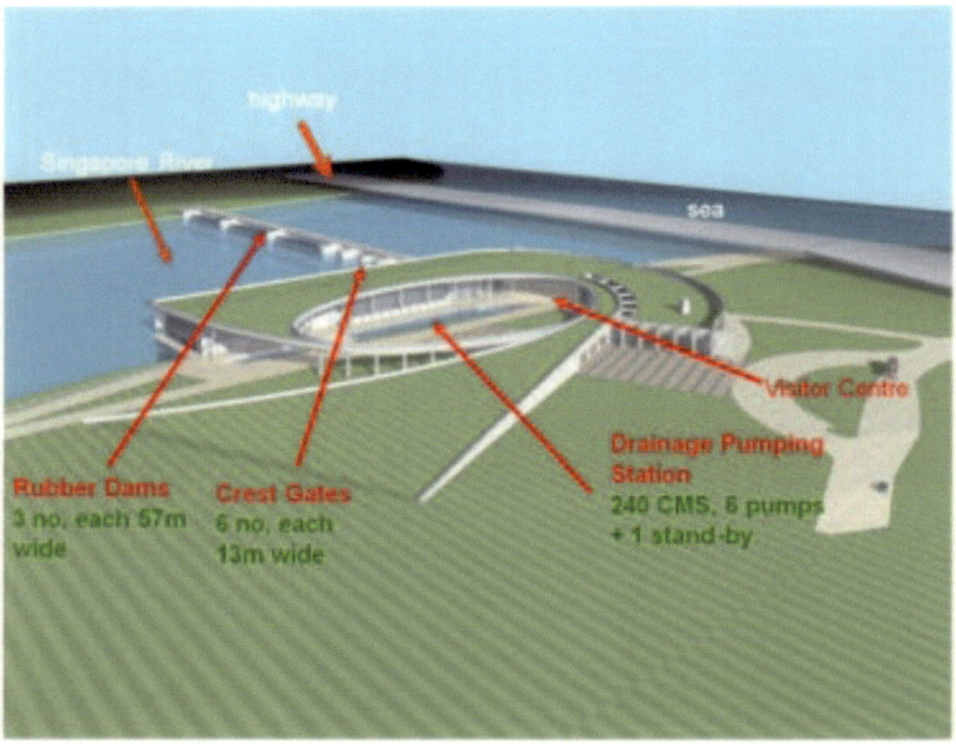

Figure 8. Model of the Marina Barrage Design

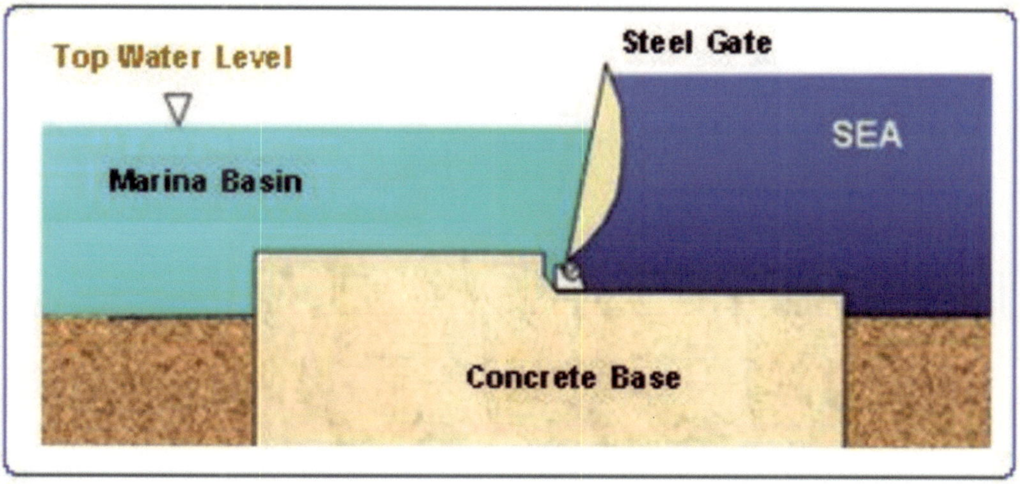

Figure 9. Working Principal of the Marina Barrage

The barrage, which comprises nine numbers of 26.8-metre-long hydraulically operated steel crest gates, will be built across the 350m wide Marina Channel to keep out sea water. Under normal conditions, the steel gates will remain closed to isolate the reservoir from the sea. During heavy rain, the steel gates will open as necessary to release excess stormwater to the sea when the tide is low. However, when it is not possible to do so during high tide, the Drainage Pumping Station capable of pumping up to 280 cubic metres per second will pump out the excess stormwater into the sea.

For water supply, the Marina Barrage have enhanced Singapore's water supply in line with Singapore's Four National Taps water supply strategy to diversify its water sources (The 4 National Taps are: local catchment, reclaimed water, desalted water and imported water). The Marina Reservoir will have the largest urban catchment of 10,000 ha among all the reservoirs. With this Project, about 60 per cent of Singapore will become catchment area.

For flood control, the Marina Barrage is part of a comprehensive flood control scheme to alleviate existing flooding in the low-lying areas of the city such as Chinatown, Boat Quay, Jalan Besar and Geylang. During heavy rain, the series of nine crest gates at the dam will be activated to release excess storm water into the sea when the tide is low. In the case of high tide, giant pumps which are capable of pumping an Olympics-size swimming pool per minute will drain excess storm water into the sea. With the Barrage and other flood-alleviation projects, flood-prone areas in Singapore will be further reduced from the current 150 ha to 85 ha, down from 3200 ha in the 1970s.Figure 9 shows the principal of how the barrage control the flood.

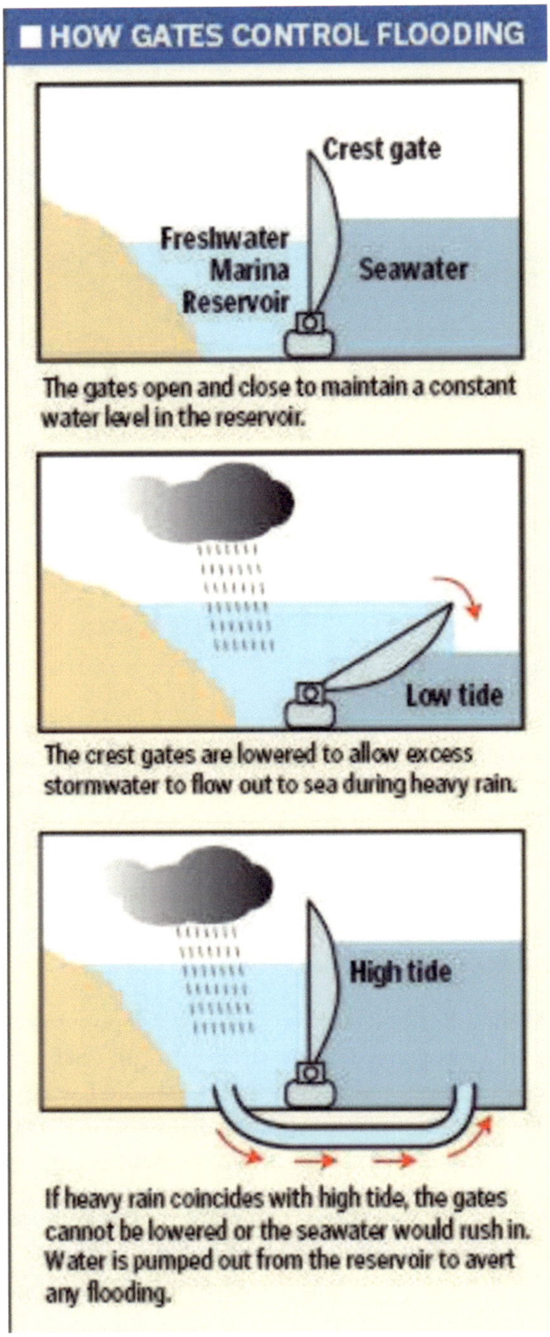

Figure 10. Principal of How the Barrage controls Flood

For lifestyle attraction, it is ideal for all kinds of recreational activities such as boating, windsurfing, kayaking and dragon boating etc. As the water in the Marina Basin is unaffected by the tides, its water level will be kept constant all year round.Marina Barrage is a showpiece of environmental sustainability, and won the Green Mark Platinum Infrastructure Award, the top award at the BCA Awards organised by the Building and Construction Authority in May 2009.

4. Environmental and social impact

The coastal reservoir at river mouth can capture every single drop of runoff, and also has the potential to collect all contaminants yielded from the catchment. For water quality, coastal reservoirs catch the stormwater runoff which is closer to drinking water in quality than the treated salt water is that is produced by desalination. Only high quality water that is free of contaminates will be allowed to enter the coastal reservoir for later use, all poor quality water will drained into the ocean guaranteeing an acceptable level of water quality.

For environmental impact, as there is nearly no land requirements for building coastal reservoirs and water quality can be guaranteed, the damage to the local ecosystems and marine wildlife can be minimized greatly. The proposed method of harvesting will avoid any severe environmental impacts such as depriving local rivers of water. As the reservoirs construction at sea, no land is needed for construction purposes making them better for the environment then typical inland mountainous reservoirs. Only high quality water (free of contaminates) will be allowed to enter the reservoir, which will help to minimize the impact of the coastal reservoir on the local ecosystem and marine wildlife.

Due to the reservoirs ability to capture every single drop of runoff it will always be functional as long as rain continues to fall. This allows for a stormwater capturing technique that is highly sustainable, will last for many decades and does not pose any major risks to the population. Coastal reservoirs will not lead to flooding, as the water collected is in the ocean. When there is too much stormwater which may occur during long wet periods, the excess water will be drained into the ocean this preventing any form of flooding. For place making, coastal reservoirs are able to fit into any location at the mouth of a river (river estuary) entering into ocean. This is very convenient for coastal cities which have huge population and less space.

5. Comparison between coastal reservoir and other water solutions

Energy use: Energy use on coastal reservoir is mainly on the transfers of potable water to areas above sea level from a water treatment plant like desalination plantthat also needs to transfer potable water to users above sea level.Water Recycling from a wastewater treatment plant needs more energy as it needs to be transported to inland reservoirs for mixing before it is distributed to users, this incurs high energy cost of mass water transport.

Emigrant cost: as coastal reservoirs are usually built near to the estuary in the sea, there is almost no emigrant cost. But inland reservoirs can affect large amounts of people especially when damming major valleys. For desalination, water recycling and mass water transport, there is no need to emigrate too many people while some infrastructure may still require property acquisition.

Waste output: coastal reservoirs and inland reservoirs have no waste output with comparison of desalination, which has high wastage from desalination (as little as 30% freshwater yields from seawater intake). Waste output of water recycling is mainly from water treatment waste.

Usable lifespan: coastal reservoirs are affected by siltation and rainfall variability, the same as inland reservoirs. As seawater will always be available, desalination mainly depends on sustainable energy supply. Water recycling which is similar to desalination, as waste water will always be available, depends on sustainable energy supply, which is similar to desalination. For mass water transport, it depends on both reliability of source and sustainable energy supply.

Maintenance cost: The maintenance cost on coastal reservoirs and inland reservoirs is low as it in mainly used for general dam maintenance, besides extra cost on coastal reservoirs includes coastal erosion / salt affects' protection. For desalination, exceptcoastal erosion / salt affects protection, the maintenance also includes treatment systems, servicing and parts replacement. So do the maintenance of water recycling. High maintenance of mass water transport cost on transport device, water transfer systems and management plans / agreements.

Potential for further capability change: coastal reservoirs' potential is high for the storage volume can be expanded at most sites, while inland reservoirs can't be changed very muchas initial design is usually for maximum possible yield. It is also not difficult for desalination to enlarge or to reduce water yield, which only needmore desalination plants can be built at the coast if energy is available. Waste water recycling is similar to desalination, which need more waste water treatment built. But potable reuse is still a distant possibility and may never be implemented except under extreme conditions [32]. For mass water transport, it depends on availability of remote sources and availability of energy.

Table 3 shows the comparison of construction cost and cost per kilolitre of water between coastal reservoir and other water solutions

	Coastal Reservoirs	Inland Reservoirs	Desalination	Water Recycling	Mass water transport
Construction cost per kilolitre of water(US$)	2.67-6.01	5.83-7.5	6.41-10.08	5.57-8.30	2.75-6.37
Cost per kilolitre of water(US$)	0.15-0.25	0.34-0.4	0.43-1.13	1.44-1.53	0.39-6.1

Table 3. Relative comparison of Coastal Reservoirs to other Australia Water Resources

The data of the table is based on the conditions in Australia. The construction cost per kilolitre of waterand cost per kilolitre of water of coastal reservoirs are calculated based on the existing coastal reservoirs in the world [33-37]. For inland reservoir, the data is based on statistical data of the UnitedNations [38]. For desalination, the construction cost per kilolitre of waterand cost per kilolitre is from comprehensive evaluation of Israel [39], Singapore [40], and Australia [41],America [42-43], India [44-45] and so on, which in the world mainly have robust desalination technology. For water recycling, the data is on account ofRemco Engineering on Water Systems and Controls in USA [46] and National Snapshot of Current andPlanned Water Recycling and ReuseRates in Australia [47]. The data resource about mass water transport is in view of water transport project in Australia [48], the USA [49-50], China [51] and Africa, which of the countries in the world are famous for water transporting. From the above, coastal reservoirs are low-cost compared with other water solutions.

6. Conclusions

"Water is everybody's business".

One of the Key Messages in the 2nd World Water Forum

Different from the proposals of desalination and wastewater recycling, the proposal of coastal reservoir will use the natural catchment runoff and there will be no need to separate the freshwater and salt or wastes, thus the energy cost associated with the treatment will be zero, there will be no carbon dioxide emission either. The pumping costs for proposals of desalination, wastewater recycling and coastal reservoir will be almost same and negligible relative to their treatment cost, thus no comparison will be given in this study. When compared with the proposal of on-land reservoirs, the method of coastal reservoirs has no cost to cover the inundation of land and people's relocation, normally this is very expensive and could be more than half of dam's construction cost. The existing coastal reservoirs in Hong Kong that have existed over 50 years, have no significant impacts on the ecosystem, and also no evidence from other coastal reservoirs in China, Korea and Singapore etc, shows that coastal reservoirs have significant environmental impacts. This conclusion could be valid for the proposed coastal reservoir in this study as fish will still have the passage to go to upstream; every year only the excessive floodwater will be diverted and the time and amount of water diversion could be adjusted to reach the win-win solution for water resources development and ecosystem like fish breeding and stocks alike. It should be stressed that experts from different disciplines should be invited to clarify the environmental impacts of coastal reservoir.

The main advantage of coastal reservoir is storing excess fresh water in the rainy season, and then the fresh water can be transferred to near watersheds through artificial channels or pipelines. Last, the fresh water can be supplied people to drink and meet the needs of agriculture [31].These do not require mountainous areas for dam construction. The highest growth in population and therefore water demand is usually expected to be close to the coast. Coastal reservoir can be located close to the high demand area.

Water demand is related to population growth, in theory population increases in coastal areas are much higher relative to the inland regions, and hence the water demand in coastal regions is likely to continue to be higher than inland regions. The future water demand in coastal regions is difficult to meet by developing more on-land reservoirs as it needs an ideal combination of suitable hydrological, geological and topographic conditions. Therefore, coastal reservoir will play a more and more important role in the freshwater resources development in the near future.

Acknowledgements

The research presented in this paper has been supported by the University of Wollongong, Australia. The works are also supported, in part, by the open fund (SKLH-OF-1002) provided by State Key Laboratory of Hydraulics and Mountain River Engineering at Sichuan University, and the National Natural Science Foundation of China (51061130547) and (51179015), (51228901) .

Author details

Shuqing Yang[1], Jianli Liu[1], Pengzhi Lin[2] and Changbo Jiang[3]

1 School of Civil, Mining& Environmental Engineering, University of Wollongong, Australia

2 State Key Lab. of Hydraulics and Mountain River Engineering, Sichuan University, China

3 College of Hydraulic Engineering, Changsha University of Science and Technology, China

References

[1] WorldWaterCouncil. Water Crisis. http://www.worldwatercouncil.org/index.php?id=25 (accessed 22 July 2012).

[2] WHO/UNICEF. Water Supply and Sanitation. 2012.

[3] WaterAmbassadors. Water Facts. http://www.waterambassadorscanada.org/water_crisis.html (accessed 22 July 2012).

[4] Lubin G. MAP OF THE DAY: The World Water Crisis. http://www.businessinsider.com/water-crisis-2011-3#ixzz1wGBzeUdf (accessed 22 July 2012).

[5] Arnell NW. Climate Change and Global Water Resources: SRES Emissions and Socio-economic Scenarios. Global Environmental Change 2004; 14(1) 31-52.

[6] Shah T, Singh O and Mukherji A. Some aspects of South Asia's Groundwater Irrigation Economy:Analyses from ASurvey in India, Pakistan, Nepal Terai and Bangladesh. Hydrogeology Journal 2006; 14(3) 286-309.

[7] Fedoroff N, Battisti D, Beachy R, Cooper P, Fischhoff D, Hodges C, Knauf V, Lobell D, Mazur B and Molden D. Radically Rethinking Agriculture for the 21st Century. Science 2010; 327(5967) 833-834.

[8] Hanjra MA and Qureshi ME. Global Water Crisis and Future Food Security in an Era of Climate Change. Food Policy 2010; 35(5) 365-377.

[9] CSIRO. Water Availability in the Murray–Darling Basin: A Report from CSIRO to the Australian Government. http://www.csiro.au/files/files/po0n.pdf (accessed 22 July 2012).

[10] UNDP. Human Development Report 2011.

[11] BluePlanetNetwork. The Keystone to Transform a Life. http://blueplanetnetwork.org/water/firststep (accessed 22 July 2012).

[12] Cabral KP, Limcuando VRL and Bontia JS. A Preliminary Study on the Capability of Burned Rice Hulls to Reduce Escherichia Coli in Drinking Water. 2007.

[13] Liu J, Temporal and Spatial Distribution of Population and Disaster Exposure Assessment in LECZ, China, in Environmental Geography. 2010, Shanghai Normal University: China. 41-42.

[14] NOAA. The U.S. Population Living in Coastal Watershed Counties. http://stateofthecoast.noaa.gov/population/welcome.html (accessed July 26, 2012).

[15] Hatton T. Population Growth and Urban Development. 2011.

[16] NationalSeaChangeTaskforce. A 10-point plan for coastal Australia: towards a sustainable future for our coast. 2010.

[17] Creel L, Ripple Effects: Population and Coastal Regions. 2003, Population Reference Bureau

[18] Chanson H. JDP. Railway Dams in Australia: Six Historical Structures. Transactions Newcomen Society 1999; 71(b) 283-304.

[19] Sadhwani JJ, Veza JM and Santana C. Case studies on environmental impact of seawater desalination. Desalination 2005; 185(1–3) 1-8.

[20] McCutcheon JR, McGinnis RL and Elimelech M. A Novel Ammonia—Carbon Dioxide Forward (direct) Osmosis Desalination Process. Desalination 2005; 174(1) 1-11.

[21] Van der Bruggen B and Vandecasteele C. Distillation vs. Membrane Filtration: Overview of Process Evolutions in Seawater Desalination. Desalination 2002; 143(3) 207-218.

[22] Lattemann S,Höpner, Thomas. Environmental Impact and Impact Assessment of Seawater Desalination. Desalination 2008; 220(1–3) 1-15.

[23] WHO. Desalination for Safe Water Supply, Guidance for the Health and Environmental Aspects Applicable to Desalination.

[24] S. Lattemann TH. Seawater Desalination—Impacts of Brine and Chemical Discharges on the Marine Environment. Desalination Publications, L'Aquila, Italy 2003.

[25] MEDRC. Assessment of the Composition of Desalination Plant Disposal Brines. 2002.

[26] Lattemann S. Protecting the Marine Environment. Seawater Desalination 2009; 273-299.

[27] UNEP. Seawater Desalination in Mediterranean Countries: Assessment of Environmental Impact and Proposed Guidelines for the Management of Brine. 2001.

[28] Bdour A.N. H, M.R., Tarawneh, Z. Perspectives on sustainable wastewater treatment technologies and reuse options in the urban areas of the Mediterranean region. Desalination 2009; 237(1-3) 162-174.

[29] Goonrey CM, Lechte P, Maheepala S, Mitchell VG and Perera BJC. Examining the Technical Feasibility of Using Stormwater as an Alternative Supply Source within an Existing Urban Area--ACase Study. Australian Journal of Water Resources 2007; 11(1) 13-29.

[30] Shu-Qing Yang SF. Coastal Reservoirs Can Harness Stormwater. Water Engineering Australia 2010 25-27.

[31] Yang S, Coastal reservoir construction. 2004.

[32] Asano T. Water From(waste) water- the dependable water resource. Water Science & Technology 2002; 45(8) 24.

[33] Liu X, Wang X, Guan X, Kong L and Su A. Study on Storage Capacity and Characteristic Water Level for Qingcaosha Reservoir [J]. Water Resources and Hydropower Engineering 2009; 7 002.

[34] Cho Do. The Evolution and Resolution of Conflicts on Saemangeum Reclamation Project. Ocean & Coastal Management 2007; 50(11) 930-944.

[35] Lie Hj, Cho Ch, Lee S, Kim Es, Koo BJ and Noh Jh. Changes in Marine Environment by a Large Coastal Development of the Saemangeum Reclamation Project in Korea. Ocean and Polar Research 2008; 30(4) 475-484.

[36] Lee SI, Kim BC and Oh HJ. Evaluation of Lake Modification Alternatives for Lake Sihwa, Korea. Environmental Management 2002; 29(1) 57-66.

[37] Onn LP. Water Management Issues in Singapore. Water in Mainland Southeast Asia. 2005; 29.

[38] UN, UNdata. 2012.

[39] Sitbon S, French-Run Water Plant Launched in Israel in European Jewish Press. 2005.

[40] Veatch B, Designed Desalination Plant Wins Global Water Distinction, in Black & Veatch. 2006.

[41] WaterTechnology. Perth Seawater Desalination Plant, Australia http://www.water-technology.net/projects/perth/ (accessed September 28th 2012).

[42] Sweet P, Desalination Gets a Serious Look, in Las Vegas Sun. 2008.

[43] Kranhold K, Water, Water, Everywhere, in the Wall Street Journal. 2008.

[44] SistlaPVSea, Low Temperature Thermal Desalination Plants, in Proceedings ofthe Eighth ISOPE Ocean Mining Symposium, Chennai, India. 2009: India.

[45] Chennai, Low Temperature Thermal Desalination Plants Mooted in the Hindu. 2007.

[46] Remco, Water Recycling Costs 2012, Remco Engineering on Water Systems and Controls.

[47] Kym Whiteoak RB, NadjaWiedemann. National Snapshotof Current and Planned Water Recycling and Reuse Rates. 2008.

[48] HWA. Piping Water From the Ord to Perth. 2003.

[49] Oksche A, Schülein R, Rutz C, Liebenhoff U, Dickson J, Müller H, Birnbaumer M and Rosenthal W. Vasopressin V2 Receptor Mutants that Cause X-linked Nephrogenic-Diabetes Insipid us: Analysis of Expression, Processing, and Function. Molecular Pharmacology 1996; 50(4) 820-828.

[50] Lavack C. Hydrological Changes in the Sierra Nevadas due to Loss in Glacierized Area. 2010.

[51] Berkoff J. China: The South-North Water Transfer Project--Is It Justified? Water Policy 2003; 5(1) 1-28.

Chapter 6

Productivity of Water in Large Rice (Paddy) Irrigation Schemes in the Upper Catchment of the Great Ruaha River Basin, Tanzania

Makarius Victor Mdemu and Theresia Francis

Additional information is available at the end of the chapter

http://dx.doi.org/10.5772/52471

1. Introduction

The concept of productivity of water (PW)[1] is increasingly becoming a cornerstone for sustainable river basin water resources management. Improving water productivity (WP), a measure of performance generally defined as the physical quantity or economic value derived from the use of a given quantity of water (Molden *et al.*, 2003), is one important strategy towards confronting future water scarcity. Increasing WP to obtain higher output or value for each drop of water used can play a key role in mitigating water scarcity (Molden *et al.*, 2001; UNDP, 2006). Global projections show that increases in WP and expansion of irrigated areas are required to account for half of the long-term increase in global water requirements for a food supply that will ensure food security of the projected 2050 population (Tropp *et al.*, 2006). Further, projected increases of WP by 30% and 60% in rain-fed and irrigated agriculture, respectively, are required to meet the demands for food security for the period 2000-2025 (Cook *et al.*, 2006; Rijsberman and Molden, 2001). WP is currently considered a more appropriate indicator of water system performance than the most widely used efficiency indicators, both classical and neo-classical (Seckler *et al.*, 2003). Under classical efficiency indicators, surface and groundwater drainage are counted as losses even when beneficially reused downstream, while neoclassical efficiency integrates water recycling into the concept of water-use efficiency (Sekler *et al.*, 2003; Xie *et al.*, 1993). Unlike irrigation efficiency indicators, WP provides more information on the amount of output that can be produced with a given amount of water (Guerra *et al.*, 1998). Also, WP can capture differences in the value of water for alternative uses

1 In this presentation, the terms "productivity of water (PW)" and "water productivity (WP)" are interchangeably used to mean the same thing.

(Wichelns, 2002). However, physical WP is not different from water-use efficiency (WUE) when expressed in terms of yield per unit amount of water consumed.

Water productivity may vary when evaluated at different spatial scales due to influencing factors such as crop choice, climatic patterns, irrigation technology and field-water management, land, and inputs including labor, fertilizer and machinery (Rosegrant *et al.*, 2002; Kijne *et al.*, 2002). Due to spatial variability in WP, several options exist for improving WP in agriculture at different scales. At plot and farm scales for example, options may involve combined research on plant physiology, agronomy and agricultural engineering that focuses on making transpiration more efficient or productive, reducing non-productive evaporation and making water application more precise and efficient (Molden *et al.*, 2003). At irrigation system and basin scales, options may include reducing non-beneficial depletion, reallocating water among uses and tapping uncommitted outflows resulting in more output per unit of water consumed (Molden *et al.*, 2003). As a result of climatic differences between locations, many options for improving WP need to be adapted to specific local conditions. However, local variation makes it difficult to upscale and downscale WP findings easily (Bouman, 2007). It is understood, though, that increased WP as a result of improved water management strategies at lower scales can result in either positive or negative linkages to WP at higher scales. For example, when low-value farm crops are supplied with the same amount of water that could supply high-value uses, overall productivity of basin supplies may be reduced when viewed in economic terms (Molden *et al.*, 2003). Irrigation supply in many cases is used for multiple purposes. Failure to account for other uses of irrigation water has resulted in undervaluation of irrigation water, and by extension, of investments in irrigation infrastructure by water managers. Understanding the value of water in its alternative uses is essential for improving WP, and guides the management and allocation of water supplies among competing users (Renwick, 2001).

The river basin is a preferred unit of analysis for water resources management. Strategies for improved water management are now typically attuned to river basin scale. Seckler *et al.* (2003) argue that, in a river basin where drainage from upstream users can be reused downstream, water losses occurring at lower scales are not true losses as long as they are or can be recovered and reused downstream. Although this argument has been useful in redefining the irrigation efficiency concept (Winchelns, 2002), which is important at basin scale in particular, it tends to downplay the importance of field-scale WP improvement interventions (Bouman, 2007). Reuse of water in many cases entails additional costs to users such as added cost to pump drain water, which many poor farmers cannot afford (Hafez, 2003; Bouman, 2007), and reductions in water quality. Improvement of WP at lower levels is also important, since it can be directly translated to improved livelihoods of farmers. Wichelns (2002) further emphasizes that improvements in farm-level water management enhance the economic values generated with limited water resources even if measures of basin WP are within desirable ranges.

Definitions and frameworks of analysis for understanding WP exist (Molden, 1997; Bouman, 2007; Molden *et al.*, 2003). Similarly, general principles underlying WP improvements and water saving at different spatial scales are elaborated (Molden *et al.*, 2003 and Cook *et al.*, 2006; Guerra *et al.*, 1998). However, these assumptions and principles are derived largely from studies conducted in Asia, and as such are not geographically applicable uniformly (Renwick, 2001; Molden *et al.*, 2001; Dong *et al.*, 2001; Singh, 2005). In particular, few such detailed studies (Mdemu *et al.*, 2004; Kadigi *et al.*, 2004; Igbadun *et al.*, 2006) have been conducted in the Great Ruaha River Basin. Since the current levels of WP in most large and small rice irrigation systems are not well understood, it is difficult to determine at which level WP can be increased from improved water management practices. A clear understanding of WP at different spatial scales was important as a precondition to implementation of any improvement strategies. The current study therefore addressed this knowledge gap by providing WP analysis of the large rice irrigation within the Usangu plains in the Great Ruaha River Basin (GRRB) in Tanzania.

2. Problem statement, rationale and research objectives

The Usangu plains in the upper part of the GRRB is important regionally and nationally due to the fact that more than 30,000 households directly depend on rice irrigation and more than 250,000 peoples indirectly depend on the rice farming from the plains. Increased abstraction of water for irrigation which is associated with poor irrigation water management practices in the Usangu plains, have resulted into significant reduction in downstream flows. As a result, conflicts between water users and uses are rife. Sustaining water allocation of compteting uses for sustainable management of water resources in the basin depends on among others decisions informed by understanding of productivity of irrigation water, both in large and small scale irrigation schemes. However, the current understanding of productivity of water to inform such decisions by water managers in the basin is inadequate. This study address this knowledge gap through assessment of water productivity of a large scale irrigation scheme from Usangu Plains in the South-western highland zone of Tanzania.

While modeling techniques can be used to assess productivity of water for different sectors over a large area within a relatively reasonable manageable time and cost, such results cannot be taken directly for policy implementation at local levels due to aggregation that obscure small but important livelihood based water uses. Therefore, the research was relevant locally because it provides an insight on the inclusive value of water at local level that can be taken into consideration during a day to day management of irrigation water resources. Once the value of water for rice plus the additional irrigaion water uses is known, then different strategies for improvements can be explored to improve socio economic activities of rural communities i.e. farmers, livestock keepers, fishers, and natural resource harvesters who depend on the availability of water in the basin. Also, sustainability of wetlands and Ruaha National Park ecosystems is vested on the sustained river flows in the basin. Regionally, the research was relevant due to the fact that the GRRB provides water to sectors of National importance and the Government is comitted to ensuring sustainable management of water in the basin is restored. This research feeds into a pool of research findings for the GRRB basin, but with significant contributions to key questions important for strategies for improving productivity of water. The main objective of this research was to determine productivity of water in large rice irrigation schemes in the GRRB. The specific objectives were: 1) to develop seasonal water balances at field and scheme levels for rice irrigation; 2) to estimate seasonal productivity of water for irrigated rice; 3) to estimate the value of water for rice production; and 4) to propose water management related strategies for improving the current productivity of water.

3. Description of the study area

3.1. Location

The research was conducted in Kapunga rice irrigation scheme in the Usangu plain. The Usangu plain, an upper catchment of the Great Ruaha River, is located in the Southwest of Tanzania between approximately latitudes 7°41′ and 9°25′ South, and longitudes 33°40′ and 35°40′ East (Figure 1). Kapunga Rice is one of the mechanized irrigation schemes which were developed by the Government in the 1990s under the management of the then National Food Corporation (NAFCO). The scheme has about 3500 hectares of mechanized rice farm. A small holder farm of about 850 hectares was developed alongside the large rice farm to carter for the need of rice small holder farmers. The two farms share the same primary irrigation canal, but have different management arrangement. This research, although covers the entire water system of these two separate farms, focused on the large rice paddy mechanized farm. The farm was privatized in 2006 to Export trading Ltd of Dar es Salaam, Tanzania.

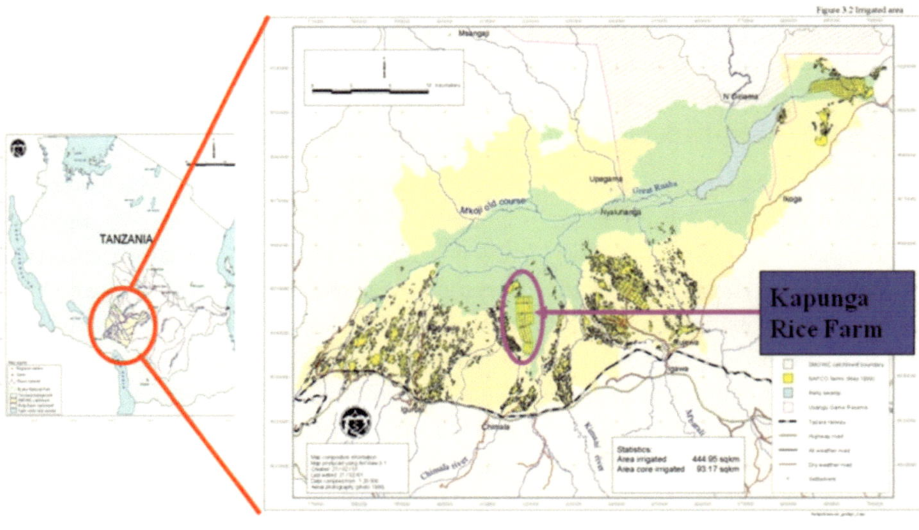

Figure 1. Location of Study Area (Adapted from SMUWC, 2001)

3.2. Climatic conditions

The general climatic pattern of the study area is tropical wet-and-dry characterized by uni-modal type of rainfall, moderate to high temperature, low wind speeds, and high relative humidity. Temperature varies between 17°C and 29°C with average mean temperature of 25°C.

The mean annual and effective rainfall received in the study area is about 669 and 479 mm, respectively. The mean relative humidity and wind speed is 62% and 2.2m/s, respectively. The rain falls between the last decads of November and April. February is the month with peak rainfall (Figure 2). The annual potential evapotranspiration is almost thrice the annual precipitation. With an exception of February, monthly evapotranspiration is higher than monthly rainfall throughout the year. This underscores the importance of irrigation water, especially for rice to facilitate crop growth, maturity and harvest.

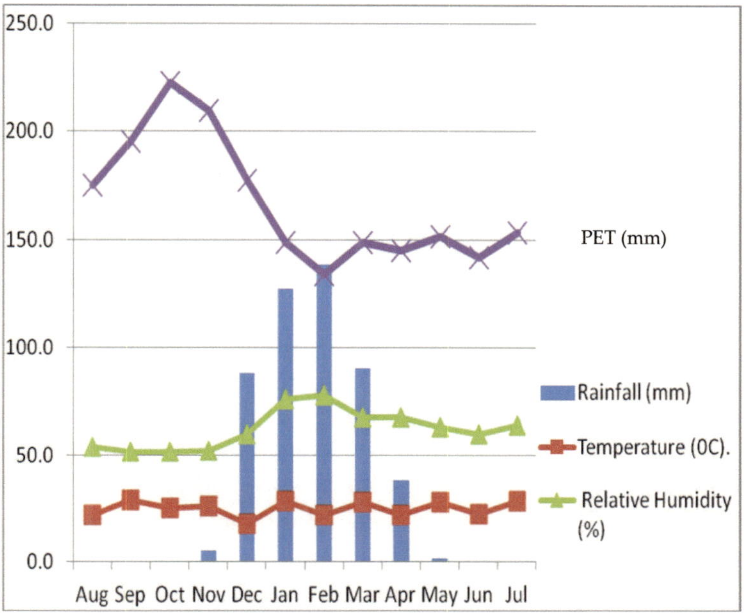

Figure 2. Long term (1980-2010) monthly average rainfall, relative humidity (RH), potential evapotranspiration (PET) and Temperature for Usangu Plain (Data Source: Kapunga Rice Farm)

3.3. Topography and hydrology

The Usangu plain is situated at 1040 meters above the sea level. The altitude on the South, South West and South East rises to above 2000 masl. A number of rivers radiates from the South and South East highlands of the plain (Figure 3). The major rivers include Ndembera, Mbarali, Ruaha, Kimani, and Chimala. These main rivers together with small rivers and tributaries provide the main source of irrigation water to all irrigation schemes in the plains.

Rivers from the upper catchments of the GRR form a life-line to the following important sectors: rainfed and dry season irrigation along the slopes of the catchments; domestic and other water use enterprises when the rivers crosses villages and towns along the Tanzania-Zambia highway; large and small scale rice irrigation in the Usangu flood plains; river rine and wetlands ecological funtions within the plains; wildlife water use in the Ruaha National Park; and hydroelectric power generation in the Mtera/Kidatu hydropower systems. From the

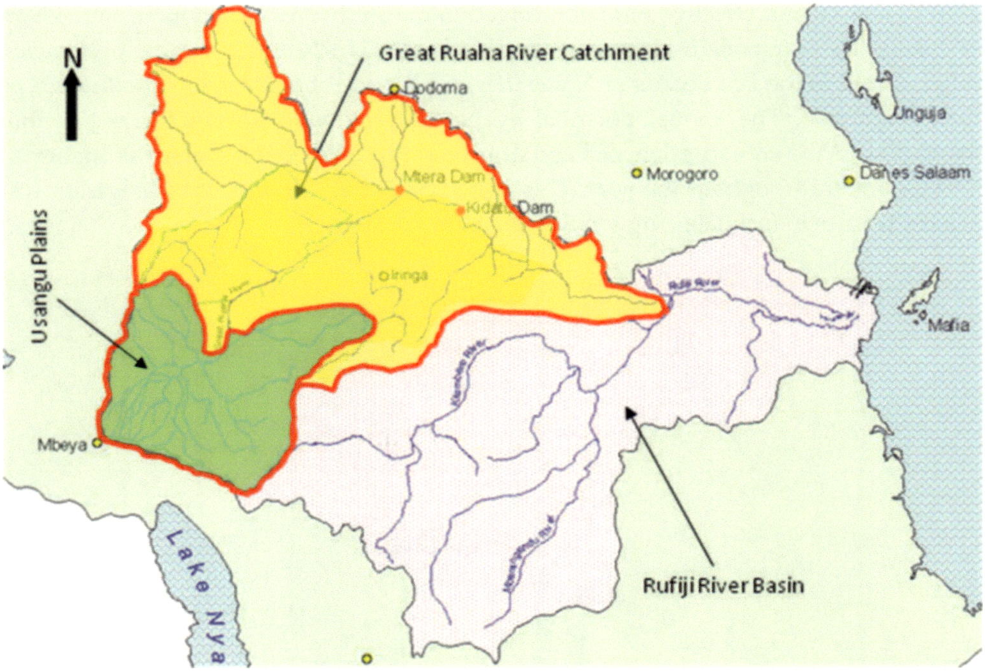

Figure 3. River drainage network, (Adapted from WWF, 2006)

Kidatu hydropower, the GRR is joined by Kilombero River and it flows into the Rufiji River. Rufiji river floods the Rufiji Delta, which is one of the most important ecological areas in Tanzania before flowing into the Indian Ocean (Figure 4).

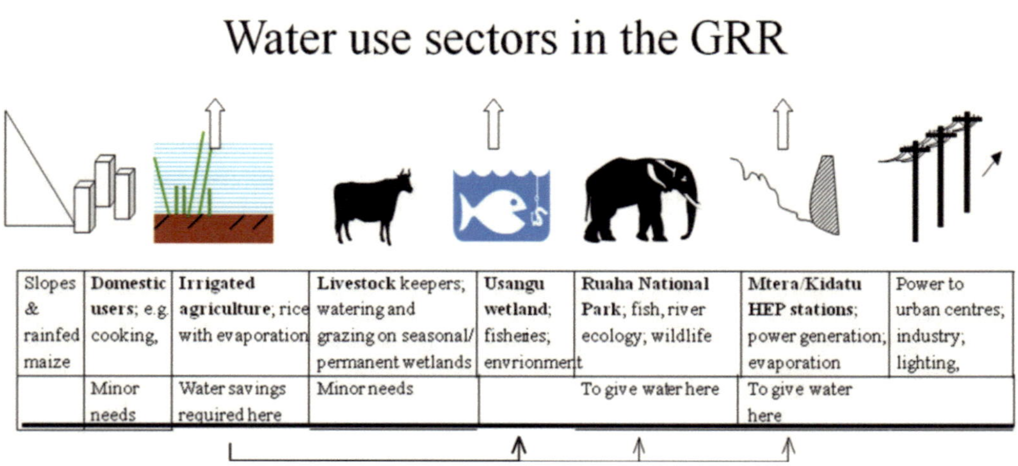

Figure 4. Water use sectors in the Great Ruaha River (Adapted from Lankford *et al.*, 2004)

4. Methodology

4.1. Research process and methods

4.1.1. Soil water balance analysis

Before water balance analysis is carried out, the irrigation system was identified and defined to consist of Kapunga large rice farms, Kapunga small holder farms and associated irrigation and drainage canals surrounding the rice irrigation farms (Figure 5). The water level gauging station in the main irrigation canal was adapted to record irrigation flows into the irrigation system. Two water level gauges were installed on the main drains of the Kapunga small holder farm and at the drainage exit from the Kapunga irrigation water system. The rating curve (water level-flow relationship equation) in the primary irrigation canal was updated through snapshot measurements of irrigation water and drainage flows using the current meter. The rating curves of the drainage canals were established and water levels in all three gauges were recorded twice per day during the period between January and December, 2010. Within the Kapunga rice farm, one rice plot was identified and monitored in terms of plot irrigation and drainage flows, standing water levels, crop season, crop yield and farm operations.

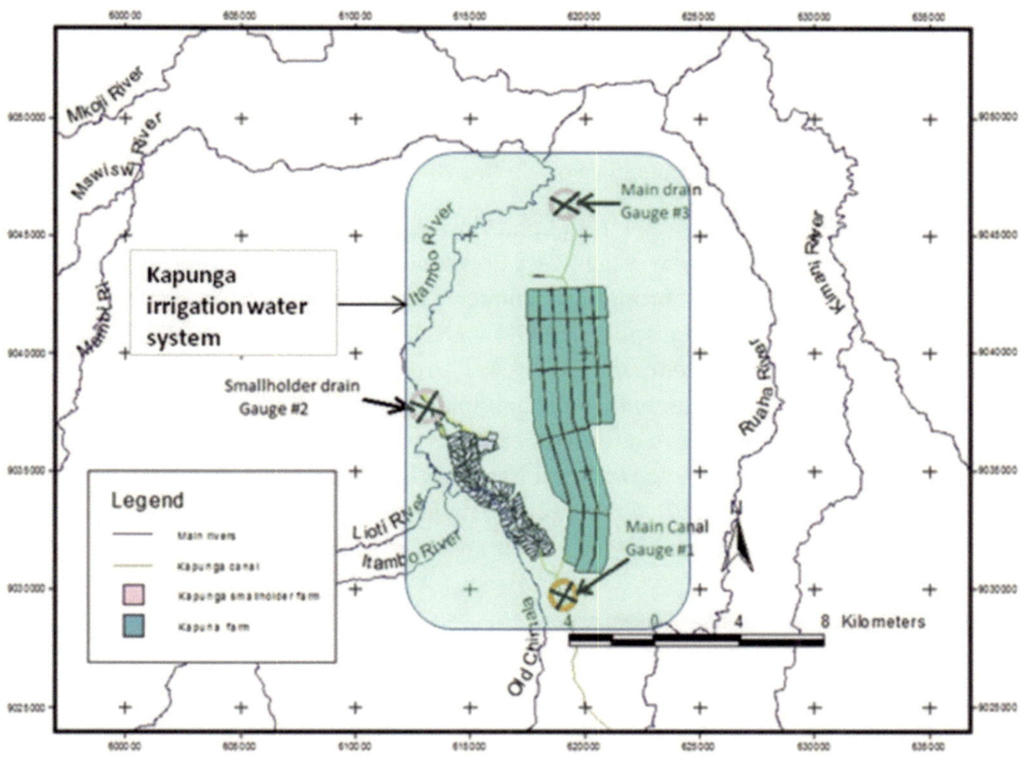

Figure 5. Kapunga Irrigation water system (Adapted from SMUWC, 2001)

Both, at irrigation system and plot levels, the soil water balance analysis was applied to partition water balance components based on equations 1 and 2:

$$Qi_{\Delta t} = Qo_{\Delta t} + \Delta S \quad (1)$$

where;

$Qi_{\Delta t}$ = total irrigation inflows (mm)

$Qo_{\Delta t}$ = total irrigation outflows (mm)

ΔS = change in soil moisture storage, +ve or -ve (mm)

$$(P + I) = (ET_c + SD + Q_{bot}) + \Delta S \quad (2)$$

where;

P = precipitation (mm),

I = irrigation (mm),

ETc = crop evapotranspiration (mm),

SD = surface drains (mm),

Q_{bot} = percolation losses (mm)

ΔS = change in soil moisture storage, +ve or -ve (mm)

The average season irrigation inflow ($Qi_{\Delta t}$) and drainage outflows (*SD*) at system and field levels were obtained from daily average water flows calculated from water level records and the rating equation established through currentmeter measurement and flow weirs installed in the sample plot. Crop evapotranspiration (ET_c) was calculated using FAO CROPWAT model with minimum and maximum temperatures (T_n & T_x), relative humidity (*RH*), wind speed (m/s), solar radiation (*R*) being the main input parameters to the model. Percolation (Q_{bot}) or ground water losses was estimated as difference in the water balance equation. Secondary soil data were collected from Uyole Agricultural Research Institute (ARI-Uyole) and used in the CROPWAT model to estimate ET_c. Change in soil moisture (Δs) was estimated based on field soil moisture content at the beginning and at the end of crop season.

4.1.2. Physical water productivity determination

Water balance components together with crop yield measurements were used to empirically determine crop water productivity as the ratio of crop yield (kg) to amount of water (available and or depleted) at field level (*Eqns. 3-5*). Three indicators were applied to estimate productivity of irrigation water. These indicators include: total inflows (*P+I*); Irrigtion water (*I*) and crop evapotranspiration (ET_c).

$$WP_{(P+I)} = \frac{Y}{(P+I)} \quad (3)$$

$$WP_{I} = \frac{Y}{I} \quad (4)$$

$$WP_{ETc} = \frac{Y}{ETc} \quad (5)$$

where;

WP = water productivity (kg/m^3)

Y = crop yield or biomass (kg)

P, I, and ETc are precipitation, irrigation water and crop evapotranspiration for the irrigated area respectively (m^3)

4.1.3. Value of irrigation water for paddy

The value of irrigation water (economic productivity of water) for paddy was estimated using a theoretical generalized profit model. In the model, each agricultural producer is assumed to maximize profit by chosing the volume of irrigation water, capital, labour, and a vector of non-water inputs for rice production. Also, producers face water constraint, reflecting their water use right or water scarcity.

For a single product of rice (paddy), *Y*, produced by the factors of production: capital (K), labor (L), other inputs (Z) and water (W), the production function can be written (Eq. 6):

$$Y = f(K, L, Z, W) \quad (6)$$

According to Euler's theorem (Chiang, 1984), the total value of product (TVP) will be exhausted if each input is paid according to its marginal productivity (VMP). Assuming competitive factor and product markets, prices may be treated as constants (constant returns to scale). By the second postulate of the residual imputation method (Eq. 7) the TVP can be written:

$$TPV_{Y} = \sum_{i=j}^{N} VMP_{i} Q_{i} \quad (7)$$

Where:

TVP is the total value product Y, VMP*i* and Q*i* are the value marginal product and quantities of resource from i=j to N. N is the number of marginal products and quantities of resources, respectively.

The first postulate of the residual method, which states that P_i = VMP_i allow replacement of P_i into (Eq. 7), which after rearranging gives (Eq. 8):

$$TVP_Y - \sum_{i=j}^{N} P_i Q_i = P_w Q_w \quad (8)$$

When all the variables in equation 6 are known, then the unknown P_w can be solved to impute the value of residual claimant (water), Pw (Eqn. 9)

$$\frac{TVP_Y - \sum_{i=j}^{N} P_i Q_i}{Qw} = P_W \quad (9)$$

A questinnaire survey was administered from a sample of rice farmers in the Kapunga large rice irrigation scheme. Inputs during the production process for different production factors were collected. The Participatory Rural Appraisal (PRA) was part of the sampling strategy using the list of rice farmers in the scheme as the main sampling framework. The survey covered a total of thirty rice farmers from the study site using semi structured questionnaire. The collected data from the questionnaire survey were used for analysis of the value of irrigation water for paddy (rice).

4.2. Results and discussions

4.2.1. Soil water balance at scheme level

For the water balance analysis, conservation of mass requires that, for the domain over the time period of interest, inflows are equal to outflows plus any change of storage within the domain. At scheme level, change in soil moisture storage between the beginning and crop harvesting is considered negligible and irrigation inflows is accounted by gross water use and surface drains. The irrigation inflows in this particular study is the amount of irrigatation water which flows for every second into Kapunga Irrigation Scheme from Ruaha river. Of the total inflows, 73% is depleted within the scheme for crop evapotranspiration, surface water evaporation, domestic uses within the scheme and groundwater losses. Twenty seven percent (27%) of the gross inflows are accounted by surface drains at the drainage exit from the Kapunga Irrigation scheme.

The amounts of water balance components at scheme level reflects the current field water use operation in the rice farms. The amount of water inflows rises in January and attains peak from February to May (Figure 6). The inflows peak months correspond to peak period of rice transplanting activities which requires sufficient amount of water for paddling of rice field, actual transplanting and maintaining standing water layer in transplanted rice fields. From May, both inflows and gross water use declines to the minimum amount (<500*l*/*s*) in July which is allowed for canal maintenance. The decline in inflows and gross water use in the indicated period coincides with rice harvesting activities and by July, almost all rice fields are completely harvested. Generally, water dependent rice farming activities in the scheme starts from October for early transplanted rice. Early transplanted rice is normally harvested starting from

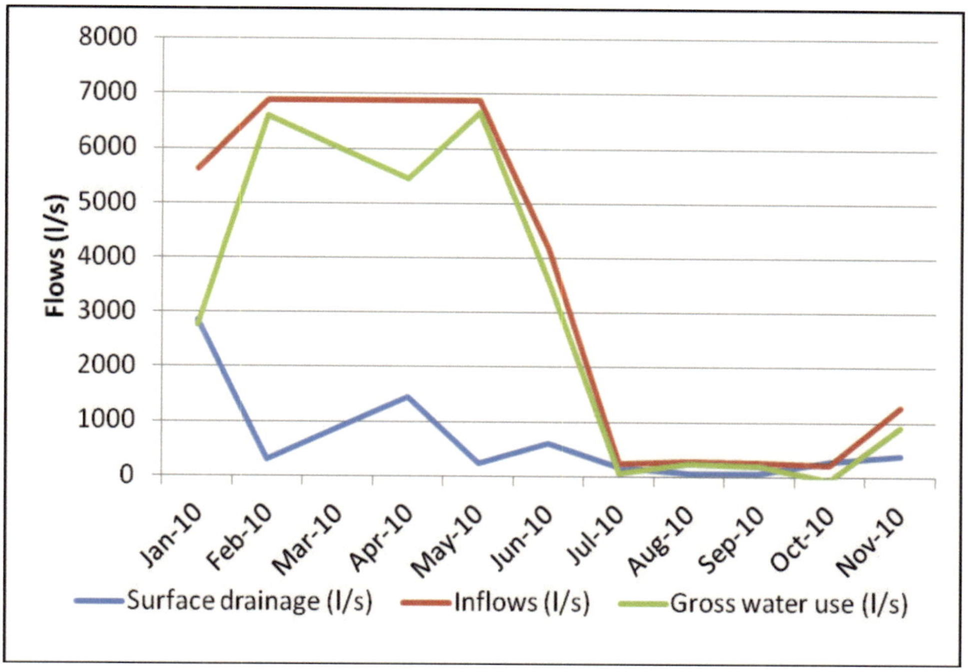

Figure 6. Irrigation inflows, gross water use and surface drainage for Kapunga water system

beginning of April while the bulk of the rice is harvested from the end of May to June. However, harvesting of late transplanted rice (April/May) may go up to July and early August.

Early transplanting has necessitated for extended rice farming season of about nine months from October to July or August, while the actual crop season of paddy rice varies between five and six months. The extended rice cropping season unnecessarily put more pressure on the limited water resources which would have been left to flow downstream for environmental uses in the wetland, Ruaha National Park and hydro-electric power generation in Mtera-Kidatu hydro system if rice crop season would have been restricted to within 6 months. For example, restricting rice crop season to six months from January would allow an extra of between 1.3m^3/s to 2.5m^3/s during November-December. This is the period when water is critically needed within the Great Ruaha River to sustain the ecosystem of the river in which portion of the river stream has been drying for the past ten years. During the 2010 crop season, a total of 3140ha were transplanted in the scheme. About 2290ha of the transplanted area belonged to Kapunga irrigation project while 850ha belonged to the Kapunga Smallholder farm. The gross irrigation inflows of 3579.21*l/s* is equivalent irrigation flows per second per hectare or irrigation hydromodule of 1.14 *l/s/ha.* The current irrigation hydromodule for Kapunga Irrigation scheme is slightly below commonly ratios (1.5-2 *l/s/ha*) used for design of irrigation systems in Tanzania. However, because of the lumped water balance for the current study, additional information would be required to determine the proportion of diverted water which is beneficially utilized for rice production at irrigation system level and ascertain performance of the scheme based irrigation flows.

4.2.2. Soil water balance at plot level

At field level, total inflows to the rice plot is partitioned into the amount of water used for rice nursery seedbed (150mm), rotavation or paddling (339 mm), transplanting (220mm) and effective rainfall- P_{eff} (480mm). Of the total inflows, 46% is lost outside the field in the form of percolation and evaporation into the atmosphere. Crop evapotranspiration (ET_c), which is only 202mm extra to effective rainfall, account for 28% of the total inflows (1). The water balance at plot level indicate that, at the end of crop season, there is a gain in soil moisture content of 34mm.

Water Balance Components	mm
Inflows (I+P)	2435.3
Nursery watering	150.0
Rotavation	339.9
Transplanting	220.0
Crop growth (up to harvesting)	1245.4
Rainfall (P)	480.0
ETc	682.7
Drains	186.5
Moisture change	-34.0
Losses (percolation, evaporation)	1120.1

Table 1. Soil water balance components at field (plot level)

Where the water losses, particularly percolation into the ground is not reclaimed for reuse within the system, then more than 54% of gross inflows at plot level is counted as water loss in the system. These water losses at plot level are influenced by use of large quantities of water during paddling, transplanting and maintaining a layer of water during the entire period of rice growth. High depth of standing water level have been observed to significantly contribute to groundwater percolation

The actual water requirement (ET_a) depends on climatic, water supply and crop growth conditions among other factors. Rice is one of the cereals with high crop water demand. Flooding, a common practice for rice farming in many rice-growing areas, increases demands for water in addition to ET_a. Rice may require over 1500 mm in tropics and sub-tropics (Guerra *et al.*, 1998) because of the flooding practices employed in rice farming. In many rice farming areas, rice fields are flooded during land preparation to facilitate plowing and paddling the fields. Paddling reduces permeability of the plow layer, significantly reducing deep percolation losses during subsequent continuous flooding after rice transplanting. Increased depth of flooded water in rice fields may, however, contribute to increased percolation losses. Other functional roles of flooding in rice fields include weed suppression resulting in reduction of herbicides application, associated costs and environmental consequences, and atmospheric cooling that reduces heat stress to crop plants. Flooding also is applied to avoid crop failure due to unexpected delays in water supplies. Incorporating these functional roles of flooding, which do not necessarily contribute to crop evapotranspiration, certainly may complicate WP assessment. Increased demands for water for rice production are therefore inherent to the farming practices. For example, in Morocco, paddy rice water use during summer varied from 1700 to 2500 mm (Lage *et al.*, 2003). However, rice ET_a may vary from about 50% and more of the total rice water requirement (2).

ET_c [mm]	Source	Location
Rice		
450-700	FAO (1986)	General
665	Lage et al. (2003)	Morocco
586-599	Mohan et al. (1996)	Sub-humid south India
640	Jehangir et al. (2004)	Sub-tropical semi arid rice-wheat zone, Pakistan
Up to 800	Ahmad et al. (2004); Singh (2005)	Semi-arid climate (Pakistan and India)

Table 2. Estimates of ET_c for rice under different climatic conditions

Therefore, ET_c may vary from one location to another due to variations in climatic conditions, availability of water supply, length of crop growth, definition of crop evapotranspiration used, and crop yield.

The consumptive crop evapotranspiration in which significant proportion of water flow in an agricultural field pass through the crop (Bouman, 2007), would theoretically lead to higher estimates of the value of water if crop yield was much higher. Although using the ETc water productivity indicator would have been economically favorable, the underlying assumption, that water supplied to crop plants will only meet crop evapotranspiration is technically not feasible under flooded surface rice irrigation schemes. Apart from groundwater percolation, evaporation from standing water surfaces constitutes a part of irrigation water requirements. Therefore, the value of water based on consumptive ET is, in general, a primary target required to be attained by water management strategies at the scale of a plot, farm and irrigation scheme.

4.2.3. Physical water productivity

Productivity of water for rice in Kapunga irrigation scheme varies between 0.17kg/m^3 and 0.62kg/m^3 (3). Using the indicators: gross inflows, irrigation and crop evapotranspiration, water productivity increases as the amount of non beneficial water uses in the water balance equation is minimized. Ideally, it is desirable to increase irrigation water productivity (WP_I) to levels close or above to Crop evapotranspiration water productivity (WP_{ETc}). Although it is not possible to eliminate the amount of water which do not contribute to crop transpiration due to its functional roles such as cooling of crop plants, significant reduction of such water is critical for increasing productivity of water in the study area. Smaller values of $WP_{(I+P)}$ and WP_I compared to WP_{ETc} is primarily due to large quantities of water used for rice field operation during the crop season.

Water use components	mm of water	Equivalent volume of water (m^3/ha)	Crop Yield (kg/ha)	Water productivity (kg/m^3)
Inflows (*I+P*)	2435.3	24353		0.17
Irrigation (*I*)	1955.3	19553	4216.67	0.22
Crop evapotranspiration (ET_c)	682.7	6827		0.62

Table 3. Water productivity of rice

Crop water productivity may show wide ranges even in comparable agro-climatic and production situations. Using the kg/m^3 ET measure from 82 literature sources, Zwart and Bastiaanssen (2005) found the range for rice to be 0.5–1.7 kg/m^3. In the analysis of water productivity data for a total of 23 irrigation systems in 11 countries in Asia, Africa and Latin America using the \$/m^3 ET water productivity indicator, the International Water Management Institute obtained values in the range from US\$ 0.03 per m^3 (for a system in India) to US\$ 0.91 per m^3 (for Burkina Faso), with an overall average of US\$ 0.25 per m^3.

High spatial variability exhibited by WP_{ETc} (4), is mainly due to crop yield and climatic variation (Tuong and Bouman, 2003). WP_{ETc} values for rice in the current study fall within the ranges of global averages (Cai and Rosegrand, 2003), but values are below the average WP_{ETc} reported by Zwart and Bastiaanssen (2003).

WP_{ETc}	Source	Location
0.4-1.60	Tuong and Bouman (2003)	Literature under Asian field conditions
0.51	Ahmad et al. (2004)	Pakistan
0.94	Singh (2005)	India
1.08	Zwart and Bastiaanssen (2003)	Review of 82 publications of the last 25 years
0.15-0.60	Cai and Rosegrant (2003)	Global averages based on 1995 production scenarios

Table 4. Water productivity of rice in terms of yield (kg) per m^3 of WP_{ETc} reported in literature

Although obtained water productivity based on gross water inflows and irrigation inflows are within WP ranges in sub-Saharan Africa (0.1-0.25kg/m^3) and other parts such as India (0.19-0.22kg/m^3) the values are low. Generally, there is potential for water productivity improvement at scheme and at plot level. Improvement in WP could be achieved through reduction of non beneficial outflows and minimizing extended crop seasons with critical implications on water use. Non beneficial water use already account for about 60% of gross water inflows on the rice plots.

5. The value of irrigation water for rice

5.1. Farming input in rice production

Apart from irrigation water, farmers need capital, non water inputs and labour to facilitate field operation during the crop season. In the current study, capital is referred to the amount of money the farmer pays to rent a 6 hectare plot, plow and rotavate the plot. The renting charge during the 2010 crop season was Tsh. one Million. After renting the plot, a number of field operations are undertaken before the paddy rice can be transplanted. The operations include clearing of the plot, plowing, preparation and raising of rice nursery. Depending on farmers circumstances, rice nurseries seedbeds can be prepared before plowing of the entire plot. When rice seedlings have reached enough height for transplanting, the field would be rotavated or paddled to facilitate for easier transplanting. All the mentioned operation requires water as an input or media to easy field operation. In the most cases, plowing and rotavating are mechanized operations.

Non water inputs in paddy rice farming include seeds, fertilizers, insecticides, herbicides sacks for bagging harvested rice (5). Seed application per hectare vary between 18.7kg and 160kg with an average application of 102.9kg/ha. Average fertilizer use during the study crop season was 156kg/ha. This amount of fertilizer is distributed into 3.7kg for rice nurseries, 87.1kg and 65.1kg during the first and second application after transplanting. Diammonium Phosphate (DAP), UREA and Sulphate of Ammonia (SA) were the common used types of fertilizers in rice farming in the study area. Herbicides (2-4D-Amine) and insecticide (*Lambda Cyhalothrin*-KARET) uses were only limited to an average of 1.5 litres and 1.2 litres per hectare respectively. Also about 32 sacks are used per hectare to fill harvested and threshed rice before they were directly sold or transported for storage.

S/N	Farm input	*N*	Minimum	Maximum	Mean
1	Seed (kg)	*31*	18.7	160.0	102.9
2	Amount of fertilizer application				
	Rice seedlings (kg)	*17*	0.7	30.0	3.7
	First application after transplanting (kg)	*29*	41.7	133.3	87.1
	Second application after transplanting (kg)	*18*	33.3	125.0	65.7
3	Herbicide (litres)	*12*	1.0	2.3	1.5
4	Insecticide (litres)	*25*	0.6	2.5	1.2
5	Sacks bagging for harvested rice	*31*	13.3	53.8	32.3

Table 5. Non water farm inputs per hectare for paddy rice in Kapunga large irrigation scheme

5.2. Labour input per hectare

A number of labour dependent activities are normally conducted during the rice crop season (6). Rice farming is one of the labour intensive activities. The average total labour requirement per hectare throughout the crop season is about 167 mandays. For rice with crop growth period of three months, the labour input is almost equivalent to two people continuously working on the farm during the entire season. Transplanting and weeding are the most labour intensive activities, each utilizing 41 and 60 mandays per crop season per hectare, respectively. Transplanting and weeding are manually done and weeds may become intensive where herbicides application is minimal and where plant spacing is high. Harvesting and bird scaring are also labor intensive activities after weeding and transplanting. Other manual field operations, i.e., farm clearing, preparation of rice nurseries, agrochemical application and bunds cleaning require less than 5mandays per hectare.

	Farm operations	N	Minimum	Maximum	Mean
1	Farm clearing	27	1.7	16.7	4.5
2	Ploughing	30	0.3	1.1	0.4
3	Preparation of rice nurseries	25	0.8	10.0	2.7
4	Rotavation	31	0.3	2.3	0.5
5	Transplanting	29	12.0	80.0	40.8
6	First weeding	27	6.3	95.0	30.4
7	Second weeding	23	9.0	60.0	29.6
8	Fertilizer application	11	0.0	0.2	0.1
9	Herbicide application	12	0.1	1.0	0.5
10	Insecticide application	12	0.0	1.0	0.3
11	Bunds clearing	30	1.0	4.5	2.1
12	Bird scaring	31	7.5	30.0	16.0
13	Canal cleaning	1	2.0	2.0	2.0
14	Harvesting	31	3.3	60.0	27.6
15	Winnowing, stitching bags and drying	6	2.7	20.0	9.1
	Total				167

Table 6. Labour use in field operation on paddy rice farm in Kapunga large irrigation scheme

5.3. Cost of farming operation, labour and farming inputs

The cost of rice farming can be divided into three main categories (7). The first category relate to direct costs of field operation, which include renting of rice plot from Export Trading Ltd, plowing and rotavation. These farm operations are mechanized and account for about 38% of the farm operation and labour cost. The second category concern the labour cost for various field operations and it accounts for 62% of the farm operation and labour cost. The third category is cost of farm inputs which include seeds and agrochemicals (fertilizer, insecticides and herbicides). Depending on the amount of fertilizer use, the total cost of farm input per hectare varies between Tsh. 180,867 and 278,167. The cost of fertilizer under this category accounts for 80% and the remaining 20% represents the cost of seeds, insecticides and herbicides. Therefore, the total cost per hectare of farming rice paddy is about Tsh. 1,332,433. For the existing 6ha plots in Kapunga Rice Project, one would need to have about eight million Tanzanian shillings in order to produce rice in the scheme.

Farming operation and labour			**Cost of farm input**			
S/N	Field activity	Cost (Tsh/ha)	S/N	Farm Input	Quantity/ha	Unit cost (Tsh)
1	Renting of the farm	166,667	1	Seeds	103kg	500/kg
2	Plowing the farm	150,000	2	Fertilizers		
3	Rotavation	83,333		• DAP	100 kg	1000/kg
				• UREA	100 kg	700/ kg
4	Nursery seedbed preparation	7,000		• SA	75 bags	700/kg
5	Farm clearing	10,000	3	Insecticides (KARET)	1 litre	2000/litre
6	Transplanting	175,000	4	Herbicides (2-4D)	1 litre	2167/litre
7	Weeding (First and Second)	133,333				
8	Labour for fertilizer application	11,000				
9	Cost of insecticide application	8,333				
10	Cost of herbicide application	8,333				
11	Bird scaring	26,267				
12	Harvesting	125,000				
13	Thrashing, winnowing and balling	150,000				
	Total	1,054,266				

Table 7. Cost of farming operation, labour and farming inputs

5.4. Crop yield, input contribution and economic return to water

Rice crop yield varies between 1493kg and 6029kg with average yield of 3517kg. The average crop yield from surveyed farmers during the 2010 crop season is less by 17% to rice yield on the sample study plot monitored during the crop season. Based on farm gate price of 600 Tsh/kg of harvested and threshed rice, the average revenue of rice production is Tsh. 2,110, 200. The revenue per hectare varies between Tsh. 895,000 and Tsh. 3, 617,400 (8).

Variable	Minimum	Maximum	Mean
Yield (kg)	1493	6029	3517
Total revenue (Tsh)	895,800	3,617,400	2,110,200
Capital (Tsh)	326,000	474,000	400,000
Labour (Tsh)	354,266	954,266	654,266
Non water inputs (Tsh)	212,167	344,167	278,167
Return to water (Tsh)	3,367	1,844,967	777,767

Table 8. Crop yield, total revenue and per hectare share of production inputs

The contribution of capital, labour and non water inputs to the average revenue stand at 19%, 31% and 13%, respectively. This implies the contribution of water to current estimated average revenue is about 37%. The contribution of water varies between 0.4% for minimum crop yield to 51% when the maximum crop yields of 6029kg/ha is attained. The estimated contribution of water to total rice production revenue shows that when capital and the value of marginal product for labour and non water inputs are higher at low crop yield in the study area, the contribution of water to the total revenue becomes much marginalized. However, the contribution of water to the total revenue is evident when the maximum crop yield is considered. The small contribution of water to the total revenue imply a low value of water for rice production.

5.5. Estimated value of water for rice production

The value of water for rice in the study area is estimated using the three water balance indicators established at plot level, i.e., gross inflows (P+I), Irrigation inflows (I) and crop evapotranspiration -ETc (9). Estimated value of water is based on the assumption that the influence of market irregularities on goods and services on farming operations during the study period was minimal and that costs were representative of local markets. The obtained value of water for rice using the three water productivity indicators is low and decreases as the volume of water considered in the productivity equation increases. This is due to the fact that there are significant proportions of non beneficial amount of water under irrigation and gross inflows.

Variable	Tsh/ha	Eq. US$/ha
Average economic return to water	777,767	519
Water balance components	Value of irrigation water	
	Tsh/m³	Eq. US$/m³
Gross inflows (P+I)	32	0.02
Irrigation inflows (I)	40	0.03
Crop evaptranspiration (ET_c)	114	0.08

Table 9. Estimated value of water for rice irrigation (The exchange rate of 1US$=1500 Tsh has been used).

The result above (9) represents typical and common characteristics of the value of water in agriculture, particularly for flood irrigated rice. Using the net value of output with and without charging family labour, Bakker and Matsuno (2001) obtained 0.03US$/m^3 and 0.05US$/m^3 as the value of water for paddy irrigated rice during the Yala season in Sri Lanka. Their estimated value increased by 75% to 0.12US$/m^3 when the gross value of output was used instead of the net value of output. However, gross value of the output includes other production inputs apart from water and therefore is not a good indicator of the value of water. In Bangladesh, the value of water for irrigated boro rice ranges from US$ 0.002 to 0.015 per m^3 (Chowdhury-undated). Rogers *et al.* (1998) estimate the value of water in agriculture in Haryana in North western India at US$ 0.02 per cubic meters. According to Sadoff *et al.*, (2003) the user value of water in irrigated agriculture is typically in the range of US$ 0.01 – US$ 0.25 per cubic meter. The lower end of this range is represented by large scale irrigation of rice and wheat while the higher end of the range is represented by high value fruits and vegetables. Comparing to municipal and other industrial water uses, the value of water for agricultural use is the lowest. Kijne *et al.*, (2003) obtained an average value US$ 0.19 per m^3 for agriculture while the value of water for industrial use was US$ 7.5 per m^3.

Hellegars (2005) argue that, where irrigation water is scarce, its marginal value ranges usually between US$ 0.05 and US$ 0.15. This value is, of course, strongly dependent on the price of agricultural products, which in turn are strongly affected by government interventions on marketing of agricultural commodities. Although the value of water estimated from the current study is within Hellegars's suggested range under irrigation scarcity, it is the farm operation, water management and price of agricultural input and outputs which determine the current estimated value of water for rice in Export Trading Ltd. This implies that there are potentials for improving the value of water through improved water management and agronomic practices in rice field operations.

6. Strategies for improving water productivity

Water related management strategies for improving WP include optimizing water use for raising rice seedlings in rice nurseries, reduction of water losses from intensive water use activities and synchronizing periods of increased water demand for rice farming activities with period of increased river flows.

Under the current practice, with an exception of Kapunga Rice Project operated plots, rice seedlings are raised in each of the 6ha plots rented by individual farmers. The number of raised nurseries for rice seedlings depends on the number of renting farmers and rented plots. Because the designs and layout of large plots lack provisions for water supply to small prepared nursery seedbeds, the entire plot is normally flooded with water when the seedlings are irrigated. As a result, about 25mm per hectare is used for water rice nurseries only. Assuming that 50% of the rice plots in Kapunga Irrigation Project are rented, then more than 40m of irrigation would be used for raising rice nurseries only. However, less water can be used if rice nurseries growing can be accommodated in few irrigation plots sparsely located within the farm. Under such arrangement, what has to be ensured is water allocation to different plots with rice seedlings.

Water intensive activities for rice production in the study area also included paddling or rotavation, transplanting and maintaining water layer throughout the crop season. Although the use of water facilitated some of the field operation, especially paddling, this softens the soil before transplanting and maintained water layer which controls weeds and therefore the cost of rice farming, increased use of water contribute to increased water losses through deep percolation and surface evaporation at plot level. For example, the average water level varied between 7cm during the first week after transplanting and 40cm during the mid of the crop season (Figure 7). Higher water levels in rice field are associated with increased deep percolation and evaporation losses.

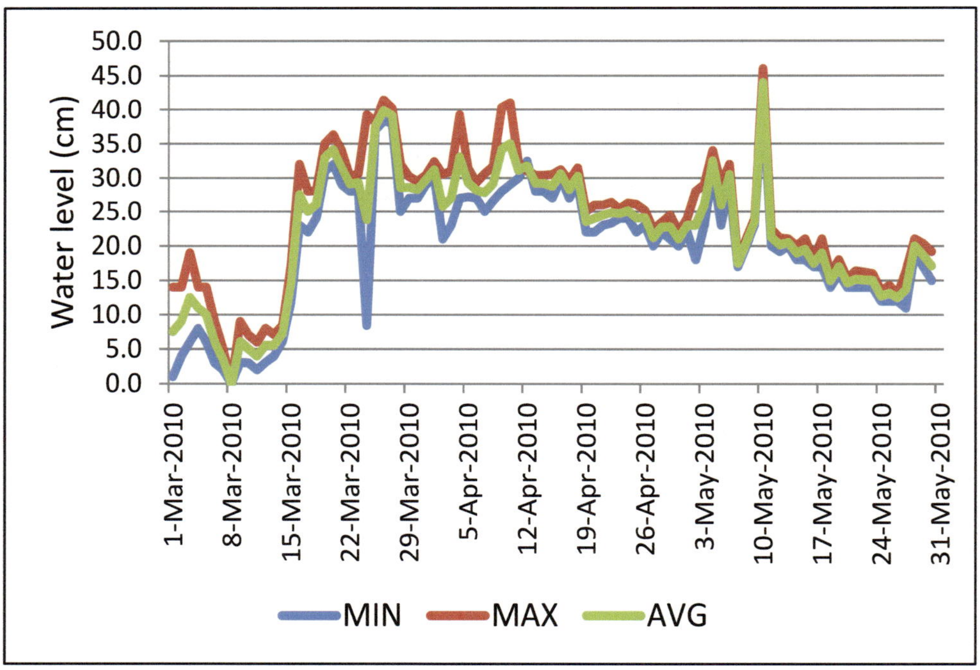

Figure 7. Water levels maintained in rice field during the crop season

Rice field preparation (farming season) in the study area normally starts between September and October each year. This is the period when river flows in the catchment including the Ruaha River which supplies water to Kapunga Rice Project are lowest. Water diversion during the period constrains the amount of environmental flows required for ecosystem services. The early start of rice farming activities prolongs crop season from the normal five months to nine months because it is only the upstream water users who get access to water at the beginning of crop season. However, if the timing for increased water demand activities (rotavation and transplanting) was synchronized with onset of rainfall and period of increased river flows (December-January) it would significantly reduce crop season and gross water use (Figure 8). For example, the effective rainfall received between mid December and end of January, which account for about 36% of the total amount of water used for land preparation, when effectively combined with irrigation water within the period, may result into reduction of gross water use to about 2100mm. The contribution of rainwater and the synchronization of peak water demand with increased river flows have not been given priority under the current management system of the Kapunga Irrigation Project.

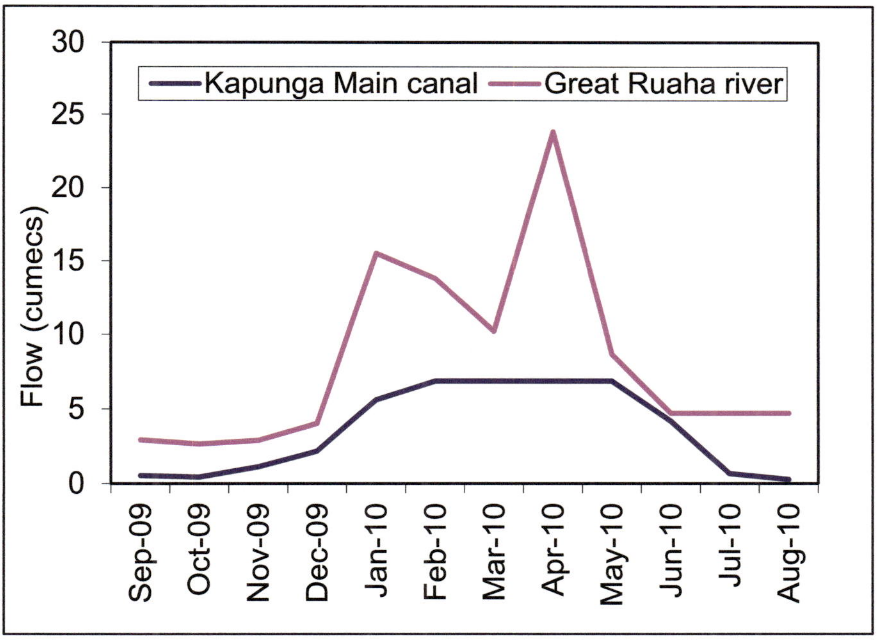

Figure 8. Seasonal hydrograph of Ruaha river and Kapunga Rice Project main canal

7. Summary and conclusion

Improving water productivity is one of the most important strategies toward confronting water scarcity. The main objective of the present study was to determine productivity of water in large rice irrigation schemes in the Great Ruaha River Basin. The specific objectives were to estimate the seasonal water balances at scheme and field levels for rice production; estimate physical productivity of water for rice; determine the value of water for rice production and propose strategies for improving productivity of water in large rice irrigation schemes.

At scheme level, 73% of the gross irrigation inflows are depleted within the irrigation scheme and the remaining 27% are counted as surface drains. However, due to lumped water balance approach in the current study, scheme depleted component of the water balance encompass groundwater and surface water losses, as well as other water use of the irrigation water. Disaggregating lumped water balance components will facilitate for precise quantification of the amount of water depleted by crops and the actual water losses into the system. At plot level, crop evapotranspiration represent only 28% of the gross inflows into the plot. Of these, effective rainfall account to 70% and the remaining 30% is provided by irrigation water. Surface drains and losses due to percolation into the ground and surface water evaporation account for 8% and 46% of the gross inflows respectively. The water balances, both at scheme and plot level, are influenced by field water dependent operation with increased volume of water use during field operation before transplanting and maintaining water layer after transplanting.

The estimated physical water productivity for rice based on gross water inflows, irrigation inflows and crop evapotranspiration vary between 0.17kg/m^3 and 0.62kg/m^3. Although obtained WP values are within the ranges of WP from different agro-ecological regions, the values are generally low, particularly for $WP_{(I+P)}$ and WP_I. However, potential for WP improvement exist through, among others, reduction of non beneficial outflows at plot and scheme levels and limiting extended rice growth period.

A number of field operations which require capital, labor and non water inputs are undertaken for rice farming. The contribution of capital, labour and non water inputs to the total revenue of rice production per hectare stand at 19%, 31% and 13%, respectively. This implies that the contribution of water to the total revenue is 37%. Based on gross inflows, irrigation inflows and crop evapotranspiration water use, estimated values of water are Tsh. 32/m^3 (US$ 0.02/m^3), Ths 42/m^3 (US$ 0.03/m^3) and Tsh 114/m^3 (US$ 0.08/m^3), respectively. The values are typical and characteristic of reported values of water in agriculture, particularly for flood irrigated rice. The obtained values of water in the current study are very much influenced by farm operations, field water management practices and the market of agricultural inputs and harvested rice product. However, as observed for estimated physical water productivity, potential also exist for improving the value of water for rice production.

A number of water management strategies can be applied to improve the current water productivity in large rice irrigation schemes. These strategies include optimisation of water use in rice field operations, reduction of water losses from rice water use activities and sychronisation periods of increased water demand for rice farming with period of increased rainfall and river flows. Such strategies, if implemented, will improve the value of water and at the same time allow sufficient flows of water to meet environmental demands in the downstream of the river.

Acknowledgements

The authors acknowledge SIDA through ARU-Sida Research Cooperation Programme for funding this research.

Author details

Makarius Victor Mdemu* and Theresia Francis

*Address all correspondence to: mmdemu@uni-bonn.de/mmdemu@aru.ac.tz

Department of Regional Development Planning, School of Urban and Regional Planning, Ardhi University, Dar es Salaam, Tanzania

References

[1] Ahmad M.D., Masih I and Turral H (2004) Diagonostic analysis of spatial and temporal variations in crop water productivity: A field scale analysis of the rice-wheat cropping systems of Punjab, Pakistan. J. Appl. Irrig. Sci. 39: 43-63

[2] Bakker, M. and Y. Matsuno 2001 A framework for valuing ecological services of irrigation water: A case of an irriagation-wetland system in Sri Lanka. Irrigation and Drainage Systemes 15:99-115

[3] Bouman, B.A.M. (2007) A conceptual framework for the improvement of crop water productivity at different spatial scales. Agric. Syst., 93 (1-3): 43-60

[4] Cai X. and Rosegrant M.W. (2003) World Water Productivity: Current Situation and Future Options. In: Kjine et al. (eds.) (2003) Water Productivity in Agriculture. CAB International, Wallingford, UK

[5] Chiang A.C. (1984) Fundamental methods of mathematical economics. Third edition. New York: McGraw-Hill

[6] Chowdhury, N.T. (undated) The economic value of water in the Ganges-Brahmaputra-Meghna (GBM) River Basin, Department of Economics, Goteborg University, Sweden

[7] Cook S., Gichuki F. and H. Turral (2006) Agricultural Water Productivity: Issues, Concepts and Approaches. Basin Focal Project Working Paper No. 1 (Draft). http://waterandfood.org/.../CPWF_Documents/Documents/Basin_Focal_Projects/BFP_restricted/Paper_1_Final_14JY06.pdf. Accessed 22 March 2012

[8] Dong B., Loeve R., Li Y.H., Cheng C.D. and Molden D. (2001) Water productivity in the Zhanghe Irrigation System: Issues of Scale. In: Barker et al. (eds) (2001) Water-

Saving Irrigation for Rice. Proceedings of an International Workshop, Wuhan, China 23-25 March 2001

[9] Guerra L.C., Bhuiyan S.I., Tuong T.P. and Barker R. (1998) Producing More Rice with Less Water from Irrigated Systems. SWIM Paper No. 5. Colombo, Sri Lanka, International Water Management Institute, 24 pp

[10] Igbadun, H.E., H.F. Mahoo, A.K.P.R. Tarimo and B.A. Salim 2006 Crop water productivity of an irrigated maize crop in Mkoji sub-catchment of the Great Ruaha River Basin, Tanzania. Agricultural water management, Vol.85 (1-2), 141-150

[11] Jehangir W., Turral H. and Masih I. (2004) Water productivity of rice crop in irrigated areas. 4th International Crop Science Congress, Brisbane, Australia. http://www.cropscience.org.au/icsc2004

[12] Kadigi, R.M.J, J.J. Kashaigili, N. S. Mdoe 2004 The economics of irrigated paddy in Usangu Basin in Tanzania: water utilization, productivity, income and livelihood implications, Physics and Chemistry of the Earth 29:1091–1100

[13] Kijne J.W., Tuong T.P., Bennett J., Bouman B. and Oweis T. (2002) Ensuring Food Security via Improvement in Crop Productivity of water. Challenge Program on Water and Food. Background Paper1.CGIAR, 44 pp

[14] Kijne, J.W., T. P. Tuong, J. Bennett, B. Bouman and T. Oweis (2003), Ensuring Food Security via Improvement in Crop Water Productivity, Challenge Program on Water and Food Background Paper 1.

[15] Kijne, J.W.; R. Barker; and D. Molden (eds.) (2003). Water Productivity in Agriculture: Limits and Opportunities for Improvement, CABI Publication, Wallingford UK and Cambridge MA USA

[16] Lage M., Bamouh A., Karrou M. and El Mourid M. (2003) Estimation of rice evapotranspiration using microlysimter technique and comparison with FAO Penman-Monteith and Pan evaporation methods under Moroccan conditions. Agron. 23: 625-631

[17] Lankford, B. A., van Koppen, B., Franks, T. and Mahoo, H. 2004. Entrenched views or insufficient science? Contested causes and solutions of water allocation; insights from the Great Ruaha River Basin, Tanzania. Agricultural Water Management 69:2 135-153

[18] Mdemu, M.V., M.D. Magayane, B. Lankford, N. Hatibu, R.M.J. Kadigi 2004 Conjoining rainfall and irrigation seasonality to enhance productivity of water in rice irrigated farms in the Upper Ruaha River Basin, Tanzania Physics and Chemistry of the Earth 29: 1119–1124

[19] Mohan S., Simhadrirao B. and Arumugam N. (1996) Comparative Study of Effective Rainfall Estimation Methods for Lowland Rice. Water Res. Manage. 10: 35-44

[20] Molden D., Murray-Rust H., Sakthivadivel R. and Makin I. (2003) A water-productivity framework for understanding and action. In: Kijne W, Barker R and Molden D (eds) Water productivity in agriculture: Limits and opportunities for improvements. CAB International, Wallingford, UK

[21] Molden D., Sakthivadivel R. and Habib Z. (2001) Basin-level use and productivity of water: Examples from South Asia. Research Report 49. Colombo, Sri Lanka: International Water Management Institute (IWMI) 24 pp

[22] Molden, D. (1997). Accounting for water use and productivity. SWIM Paper 1. Colombo, Sri Lanka: International Irrigation Management Institute. 26pp.

[23] Renwick M.E. (2001) Valuing water in a multiple-use system: Irrigated agriculture and reservoir fisheries. Irrig. Drainage Syst. 15: 149–171

[24] Rijsberman F.R. and Molden D. (2001) Balancing water uses: water for food and water for nature, thematic background paper. International conference on freshwater, Bonn, 18 pp

[25] Rogers, P., R. Bhatia and A. Huber (1998), Water as a Social and Economic Good: How to Put the Principle into Practice, paper prepared for the meeting of the Technical Advisory Committee of the Global Water Partnership in Namibia.

[26] Rosegrant M.W., Cai X. and Cline S.A. (2002) World Water and Food to 2025, Dealing with Scarcity. International Food Policy Research Institute, Washington D.C. 322p

[27] Sadoff, C., D. Whittington and D. Grey (2003), Africa's International Rivers: An Economic Perspective, World Bank, Washington DC.

[28] Seckler D., Molden D. and Sakthivadivel R. (2003) The Concept of Efficiency in Water resources Management and Policy. In: Kijne et al. (eds.) (2003) Water Productivity in Agriculture. CAB International, Wallingford, UK

[29] Singh R. (2005) Water productivity analysis from field to regional scale: integration of crop and soil modeling, remote sensing and geographical information. Doctoral thesis, Wageningen University, Wageningen, The Netherlands, 146 pp

[30] SMWUC, (2001) Main Report. The Usangu catchment Baseline Report.

[31] Tropp H., Falkenmark M. and Lundqvist J. (2006) Water governance challenges: Managing competition and scarcity for hunger and poverty reduction and environmental sustainability. Background paper (Final draft): Human Development Report 2006, Water for Human Development, 46 pp

[32] Tuong T.P. and Bouman B.A.M. (2003) Rice Production in Water-scarce Environments. In: Kjine et al. (eds.) (2003) Water Productivity in Agriculture. CAB International, Wallingford, UK

[33] UNDP (2006) Beyond scarcity: Power, poverty and the global water crisis. Human Development Report 2006, 422 pp

[34] Wichelns D. (2002) An economic perspective on the potential gains from improvement in irrigation water management. Agric. Water Manage. 52: 233-248

[35] WWF, 2006 Integrated water resource management in the Great Ruaha River Catchment, Tanzania, Project Proposal

[36] Xie M., Küffner U. and Le Moigne G. (1993) Using Water Efficiently. Technological Options. World Bank Technical Paper No. 205. Washington D.C., World Bank

[37] Zwart S.J. and Bastiaanssen W.G.M. (2003) Review of measured crop water productivity values for irrigated wheat, rice, cotton and maize. Agric. Water Manage. 69(2): 115-133

Chapter 7

Location of the Rainfall Recharge Areas in the Basin of La Paz, BCS, México

Arturo Cruz-Falcón, Enrique Troyo-Diéguez,
Héctor Fraga-Palomino and Juan Vega-Mayagoitia

Additional information is available at the end of the chapter

http://dx.doi.org/10.5772/52518

1. Introduction

Rains are very scarce in the state of Baja California Sur, México, with an annual average of 175 mm (SARH-CNA, 1991). The basin of La Paz records a mean annual precipitation of 265 mm/year (CNA, 2005). Rainfall occur in torrential of short duration during the cyclone season (late summer and early autumn), which not only causes flooding, but because a large amount of water falls in a short time, it drains quickly towards the sea. Anyhow, infiltration is carried out, allowing aquifer recharge. Even so, available water for the population and agricultural irrigation is groundwater, which is pumped from the aquifer.

Due to an increased water demand, water availability from aquifers in the state of Baja California Sur, has been decreasing rapidly, mostly because of an inadequate management as it pumps more water than the supplied by natural recharge. This is the reason why most aquifers, in the state, are over-exploited.

An example of over-exploitation takes place in the agricultural valley of Santo Domingo, located about 120 miles North of La Paz, but now this serious situation is about to happen in the aquifer of La Paz.

For the sustainability of these aquifers, it is necessary, among other things, to handle a strict management program (that does not exist at present). It is necessary to maintain a balance between recharge and exploitation, and also to protect recharge areas to avoid reducing the volume of rainwater that infiltrates into the ground.

Groundwater recharge is a process by which surface water or rainwater percolates through the soil to the level of groundwater to supply water for aquifers (Davis and Wiest, 1971).

This recharge is essential to basins and aquifers for water storage, especially in regions of high water demand, in arid zones without rivers or lakes, and where water is essential to economic development.

The city of La Paz is located at the Southern tip of the peninsula of Baja California (Figure 1). For years the source of freshwater supply for the population and agriculture has been obtained from groundwater exploitation of the aquifer of La Paz, but in recent years due to increased water demand by population growth, the aquifer is in over-exploitation condition (CNA, 1997).

The basin of La Paz collects 410 Mm^3/year of rainfall, from this, evapotranspiration is about 80%, 16% infiltrates into the ground for aquifer recharge, and 15 Mm^3/year runoff to the sea (Cruz-Falcón et al., 2011).

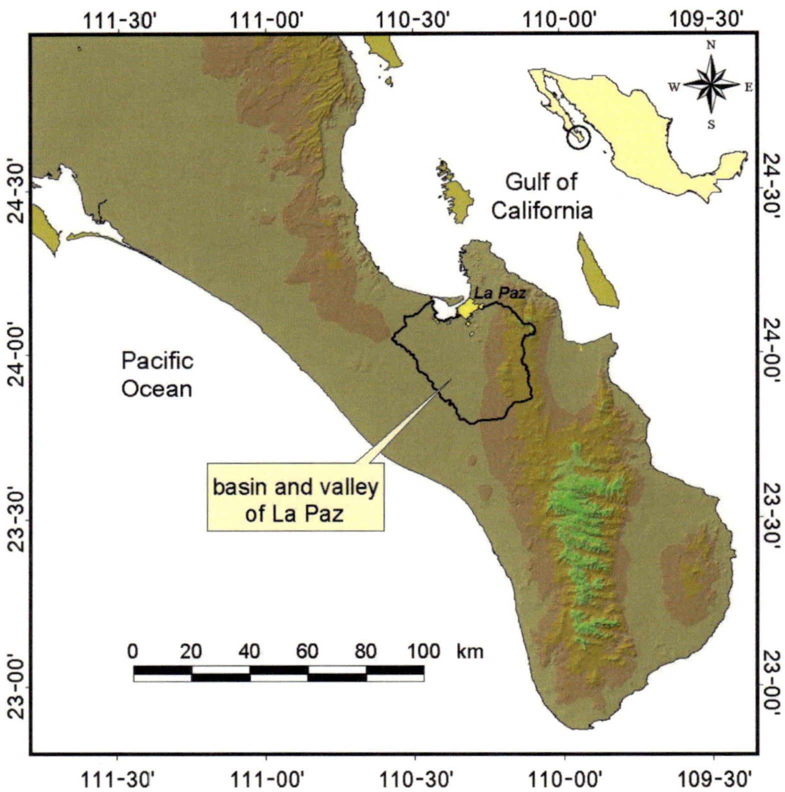

Figure 1. Location of the study area

For estimating groundwater recharge, literature describes several methods. However, these methods are not clearly defined and ordered (Sophocleous, 1991; Kommadath, 2000, Bridget et al., 2002), but can be grouped in two general classes, potential and real methods. Generally each method is related to another, so none is completely independent.

http://dx.doi.org/10.5772/52518

To choose an appropriate method, it is highly dependent on climatological information and water table. In the best case, depending on existing information and field work, the most suitable is a combination of potential and real methods (Bridget et al., 2002).

Estimation of recharge in a hydrological basin or an aquifer involves some complexity, due to the physical features of the area, availability of information, quality of data, and the method used.

The process of groundwater recharge consists of water flow from surface to subsurface, which is basically determined by infiltration capacity of the sediment layers and rocks. Infiltration depends on several factors such as the permeability of the materials, soil texture, moisture content, vegetation cover, land use, air entrainment, fine material washing, and soil compaction (Aparicio-Mijares, 1992). Also the climate, rain, slope, type of rock and sediment, play an important role in this process.

Normally groundwater recharge calculation is done by taking water level data and the water table fluctuation, supported by aquifer parameters such as transmissibility and storage coefficients.

A system of any hydrological basin can be described by a balance which considers water inlets and outlets. The main component of this system is the precipitation, but because all other components are related to each other, groundwater recharge, that can be vertical (from rainfall) or horizontal (from groundwater flow) is essential for aquifers supply (Peña-Haro and Arcos-Hernández, 2004).

In the basin of La Paz, rainfall recharge is the most important inlet component, although there is a small agricultural irrigation recharge, and a little artificial recharge caused by a dam located Southeast of the city of La Paz.

Because vertical recharge normally occurs in a stratified soil where hydraulic conductivity is variable, infiltration will depend on soil constitution, so it will be different if soil is formed of gravel, sand or clay.

For rainwater to infiltrate soil surface, soil must have some favorable conditions for this to happen. For example, an area devoid of vegetation will be freely exposed to direct impact of raindrops. This can substantially compact the soil and introduce simultaneously tiny particles inside the cracks and open ducts, which result in an immediate reduction of infiltration. By contrast, a dense covering of vegetation protects the soil surface allowing a better infiltration (Davis and Wiest, 1971).

Plant roots also help to keep soil open, thereby increasing infiltration. Infiltration occurs easily over unconsolidated soil or sediment, with coarse texture, especially if they are covered with vegetation. In consolidated soil, infiltration occurs if it is degraded by environment, or in areas where fractures or faults are present (Davis and Wiest, 1971).

In some of geohydrological studies conducted in the valley of La Paz, it is mentioned about the aquifer recharge areas. Most agree that these areas are located to the East and South of the basin, but none of these studies show them geographically on the map.

The SARH-UNAM-UABCS (1986), define a rainfall area in the Eastern portion of the valley of La Paz, a recharge area at the middle of the valley, and a rainfall water collecting - infiltration area Westward. They mention that groundwater recharge comes mainly from the Valley of El Coyote (North), from the elevations on the Western side the basin, and from the Southern portion of Sierra El Novillo (Southeast). The IPN-CICIMAR/CIBNOR/UABCS (2002), reports that the main aquifer groundwater recharge comes from the East side of the valley of La Paz, as a result of infiltration of surface runoff. CIGSA (2001), mention that the flow pattern determines that recharge areas are located at the West and South of the valley of La Paz.

A technique for recharge estimation (qualitatively), using some of the physical parameters involved in groundwater recharge, in a simple and practical way, is the method of 'weights'. This method can handle the parameters and some physical factors according to the characteristics of each area, by judgment and experience of the researcher, what distinguishes its applicability in the study of surface water and groundwater.

This study identifies the potential rainfall recharge areas in the basin of La Paz, with a geographic information system (GIS) from physical parameters such as terrain slope, geology, geomorphology, soils, land use and vegetation, runoff and precipitation. Through spatial analysis raster models are generated. The classes of each model are reclassified, and weights are assigned, using a criterion of logical reasoning where a greater or lesser number is assigned according to the importance of each class in contribution to groundwater recharge.

2. Study area

The basin of La Paz is located in the Southern portion of the peninsula of Baja California, which the city of La Paz is settled (Figure 1). The climate in this area is predominantly dry with an average temperature of 20-26ºC, with higher temperatures in the months of July, August and September, sometimes reaching 40 to 45ºC. The average annual rainfall in the basin is 265 mm (CNA, 2005), which occurs mostly during the summer, with the highest precipitations in August and September. In late summer occur tropical storms and cyclones. These events bring heavy rainfall, that contribute to recharge the aquifers all over around.

2.1. Basin of La Paz

From a physical point of view, a hydrological basin is defined as a surface in form of depression that collects rainfall water. Part of the rain that falls evapotranspires, another part infiltrates into the ground, and the remaining, once the subsurface is saturated, runoff to the plains, which may be a lake, river or the sea (modified of SEMARNAT, 2001).The delimitation of the basin of La Paz used in this study was defined by Cruz-Falcón et al. (2011). It extends approximately in an area of 1,275 km^2, located between 23° 47' 24" to 24° 10' 12" North and 110° 04' 48" to 110° 35' 12" West (Figure 1). All runoffs end up into a coastal lagoon known as Lagoon of La Paz (Figure 2).

http://dx.doi.org/10.5772/52518

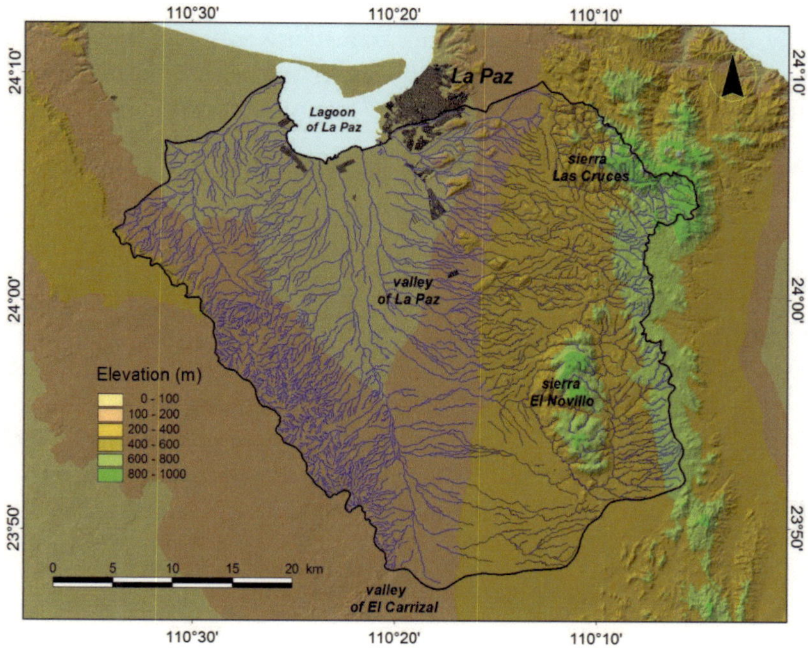

Figure 2. Delimitation of the basin of La Paz, and terrain elevation model (TEM).

3. Methodology and results

To locate potential rainfall recharge areas in the basin of La Paz, it was used a geographic information system (GIS) fed with vector information of physical parameters such as topographic slope, geology, geomorphology, soils, land use and vegetation, and precipitation. By spatial analysis, these layers of information were converted to images with a spatial resolution of 100 m.

To identify and locate rainfall recharge areas, it was used the 'weight average method' (Katpatal and Dube, 2004), which can be just called 'method of weights'.

The classes of each model were reclassified, and weights were assigned to each of them. The criteria for assigning weights to each class were based on a logical reasoning approach that considers weight a greater or lesser number according to the importance of each class as a contribution to groundwater recharge (Katpatal and Dube, 2004).

The assignment of weights to the different classes is based on an arbitrary scale, but considers the importance of each class as a contribution to groundwater recharge. The higher the number assigned, greater the contribution to groundwater recharge, and vice versa. In some cases, the same number could be assigned to different classes since their contribution to recharge is considered the same.

Once weights were assigned to the classes of the different parameters, it was generated a new model. Finally, the new models of the different parameters were integrated to generate a representative model of rainfall recharge areas in the basin of La Paz. This model contains seven classes that represent from a poor (1) to very good (7) contribution to groundwater recharge.

Weights assigned to the classes of each parameter are shown in the corresponding models (Figures).

3.1. Data used

Data come from 1983 available Geology mapping Charts from INEGI (National Institute of Statistics, Geography and Informatics) 1:250,000 scale. Soil, Land Use and Vegetation mapping Charts (2000), from INEGI,1:250,000 scale. Geomorphology map (2004), from CONABIO (National Commission for Knowledge and Use of Biodiversity) 1:1,000,000 scale. Elevation contours and runoff (1998-2003) to generate elevation model, come from Charts of INEGI, 1:50,000 scale. Precipitation model comes from Cruz-Falcón et al. (2011).

For all processes, program ArcView 3.2 was used.

3.2. Criteria and models

From the terrain elevation model (TEM) (Figure 2), generated with ArcView tool and contour mapping of INEGI (1998-2003), it was obtained the terrain slope model (Figure 3). The terrain slope is the angle in degrees from horizontal. For this model there were considered four classes, of which the smallest angle has a greater contribution to recharge and vice versa.

From Figure 3, is observed that terrain slope with an angle of 0 to 10 degrees covers most of the basin.

For the geology model (Figure 4), it was considered the type and physical condition of rock. Rocks located near the coast, and consolidated volcanic material was assigned a smaller number. For consolidated sedimentary materials was assigned a slightly larger number. To metamorphic, weathered and fractured intrusive rocks, as well as for alluvial deposits, was assigned a number of medium to high.

Álvarez et al. (1997), describes the geology of the basin of La Paz characterized by a sequence of marine sediments and alluvial recent fans formed by conglomerates, sandstones and shales. Towards the NNW of the valley the Middle Tertiary to Recent sequence is formed by: San Gregorio Formation (Late Oligocene–Early Miocene) with analternation of tuffysandstones, siltyshales, mudstones, conglomeratics and stones and fosforite inter-bedded layers; San Isidro Formation (Early Miocene) formed by glauconitic sandstones, conglomerates, shales and some pink colored rhyolitic tufflayers; and Comondu Formation (Late Miocene) formed by sandstones and volcanic conglomerates, rhyolitictuffs, and esiticlahars and lava flows. Towards the NNE, emerge rocks from Cretaceous to Recent that constitute the sierras Las Cruces and El Novillo (Figure 4), formed by intrusive rocks such as granite, gabbro and tonalite.

http://dx.doi.org/10.5772/52518

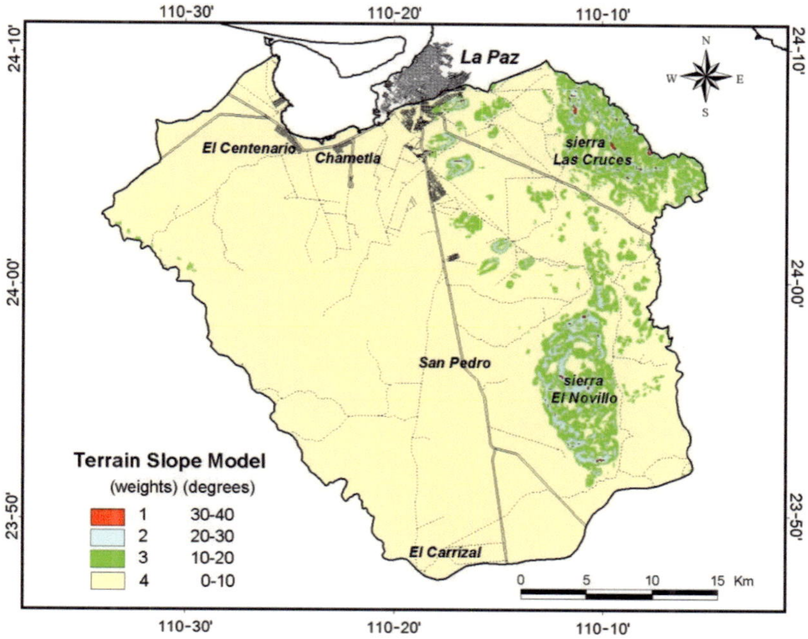

Figure 3. Terrain Slope model of the basin of La Paz.

In this study, due to the physical condition of igneous rocks as granite (KGr) that is quite fractured and weathered, it was assigned a higher number than for alluvial deposits (Qal). Gabbro (KGa) also was considered fractured and weathered.

From Figure 4, is observed that Quaternary material such as sandstone (ar), conglomerate (cg) and alluvium (al), are distributed in the middle of the basin, and metamorphic rocks (gneiss and schist) and igneous rocks (gabbro and granite) are located to the East and Southeast.

To assign weights to the classes of the geomorphology model (Figure 5), there were considered the hydro geomorphological features of the basin. Valleys (piedmont and plains) with gentle slope and low drainage were assigned with high infiltration rate, compared with mountains and hills with regular infiltration rate, and structural terraces with lower infiltration rate.

From Figure 5, is observed that most of the basin is characterized by piedmont and cumulative plains, except to the East and Northeast that is defined by mountains and hills.

For the land use and vegetation model (Figure 6), weights were assigned from two criteria, one that assumes vegetation can retain more rainwater for allowing infiltration, and the other, that urban areas and man-made infrastructure is where runoff predominates. Eventually, for high and dense vegetation was assigned a higher number than for small vegetation, as well as grassland and halophyte plants. Urban areas were assigned with lower number because there is no infiltration.

From Figure 6, is observed that in most of the basin predominates sarco-crasicaule and sarcocaule bush, except to the East and Southeast that is characterized by low forest.

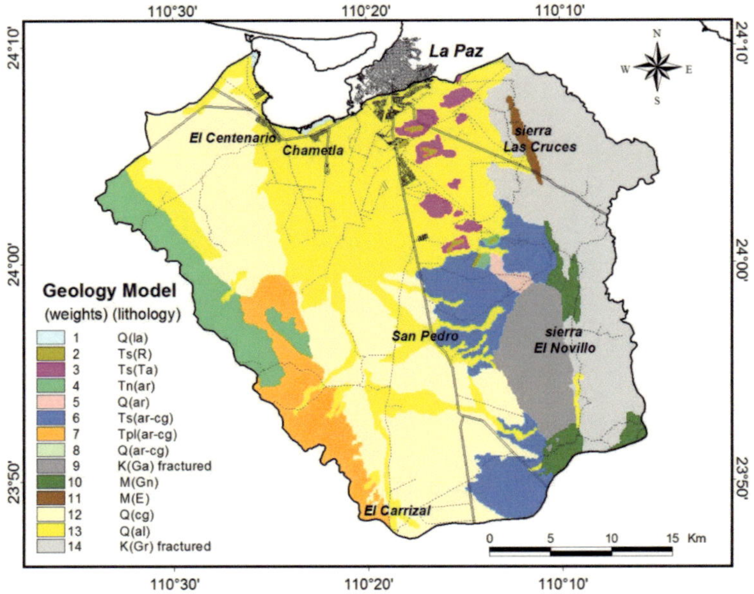

Figure 4. Geology model of the basin of La Paz.

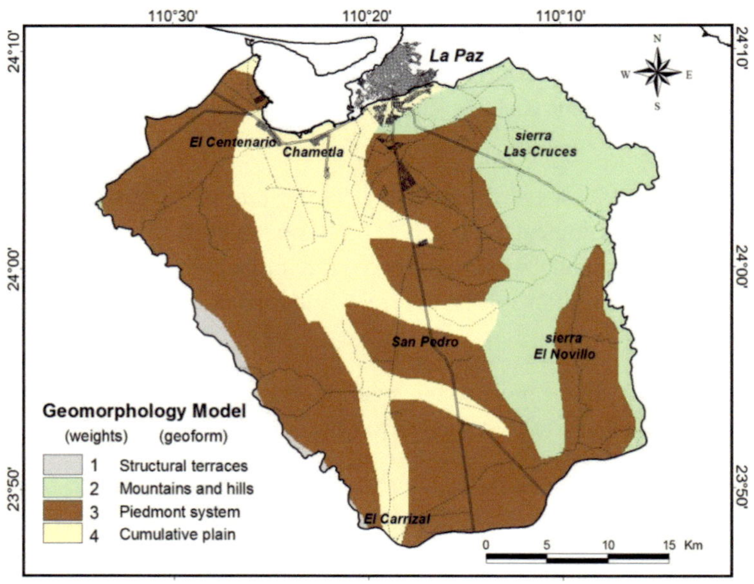

Figure 5. Geomorphology model of the basin of La Paz. Intermittent streams are shown (blue lines).

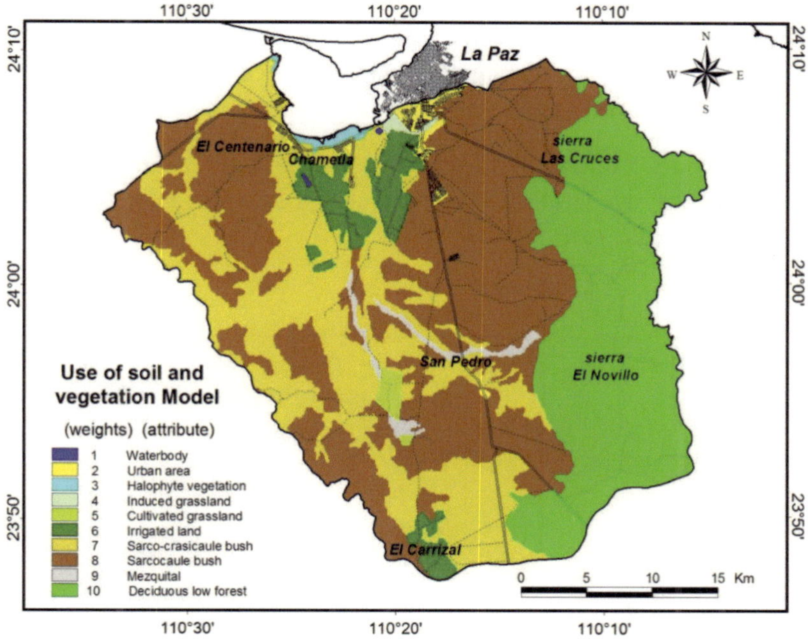

Figure 6. Use of Soil and Vegetation model of the basin of La Paz.

For the edaphic model (Figure 7) there were considered the different types of soil for the assigning of weights.

Regosol is the soil which was assigned a higher number. This soil is common in desert and dry tropical areas, it is the most abundant soil of the alluvial deposits that fill the basin. Litosol, despite of being a soil forming a very thin layer, it allows a rapid infiltration to subsurface layers. Fluvisol, usually common in alluvial deposits, has a very fine texture that diminishes infiltration capacity. Solonchack, is a predominantly saline soil that extends very close to the coast of the lagoon of La Paz. Xerosol, is a very fine soil rich in clays, that because of its texture, infiltration is reduced.

From Figure 7, is observed that most of the basin is characterized by regosol. Litosol extends Eastward, fluvisol at the middle, and xerosol to the West.

Due to the characteristics and properties of the models and the values assigned to their classes in relation to their contribution to groundwater recharge, it is observed that areas with better chance of recharge are located in plain areas, mainly alluvial. But if we assume that there is little rain in these areas, their contribution to recharge will not be the same. There for, it was necessary to include the parameter of rainfall, which is determinant in this process.

The precipitation model of the basin of La Paz (Figure 8) comes from Cruz-Falcón et al. (2011). It was generated from 25 year precipitation data (1980 to 2004), of 12 weather stations located within and outside the basin of La Paz, provided by the State National Water Commission (Comisión Nacional del Agua, CNA). To obtain this model, it was calculated the total annual precipitation (TAP) for each climatological station, and TAP of 25 years was averaged. The average TAP for each station was interpolated to obtain isohyets, using spline method. Averaged TAP contours were rasterized with a 100 m spatial resolution.

According to the model, precipitation extends in a range from 150 to 400 mm/year, where most precipitation occurs in elevated areas located to the East and SSE of the basin, but decreases toward the opposite side (NNW).

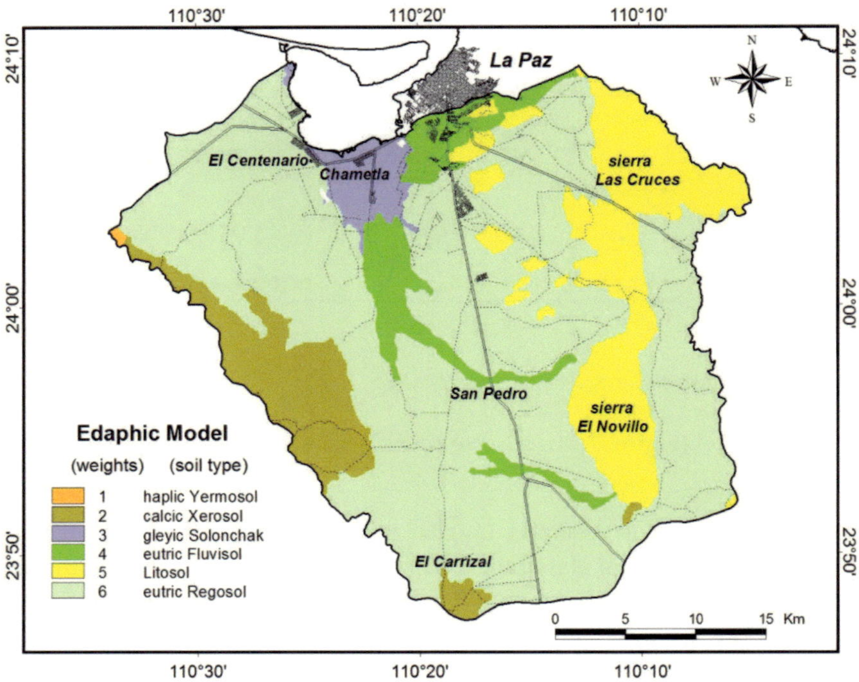

Figure 7. Edaphic model of the basin of La Paz

3.3. Integration of models

The obtained models of terrain slope, geology, geomorphology, land use and vegetation, soil and precipitation were integrated to build a new model containing the areas (polygons) of potential rainfall recharge in the basin of La Paz (Figure 9). This model includes the distribution of polygons for seven classes, from a poor (1) to very good (7) groundwater recharge.

According to the obtained results, the recharge areas that resulted from the classes: moderate (5) to very good (7), may be considered as main recharge areas. Therefore, it is assumed that rainfall potential recharge areas in the basin of La Paz, are located on the East, South and Southeast of the basin, from sierra Las Cruces to sierra El Novillo, and the Northern portion of the valley of El Carrizal (Figure 9).

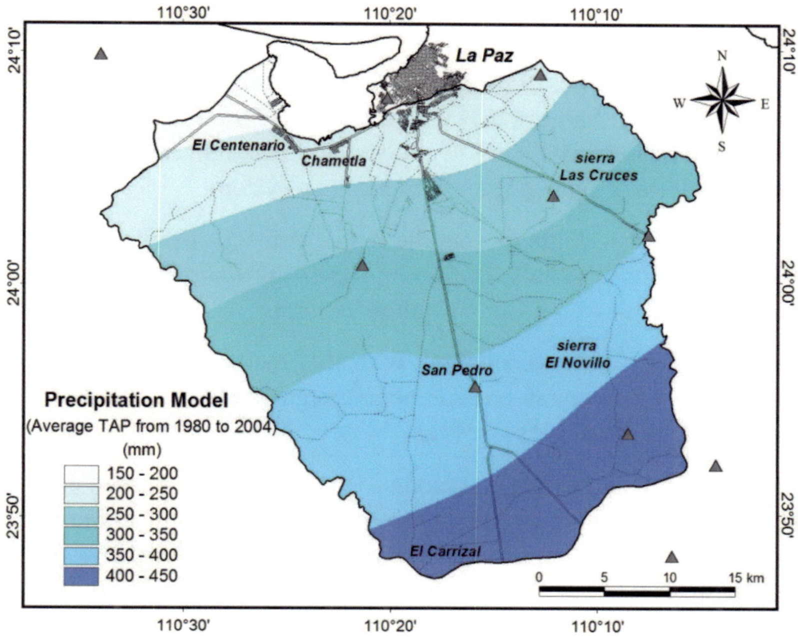

Figure 8. Precipitation model of the basin of La Paz. From Cruz-Falcón (2011).

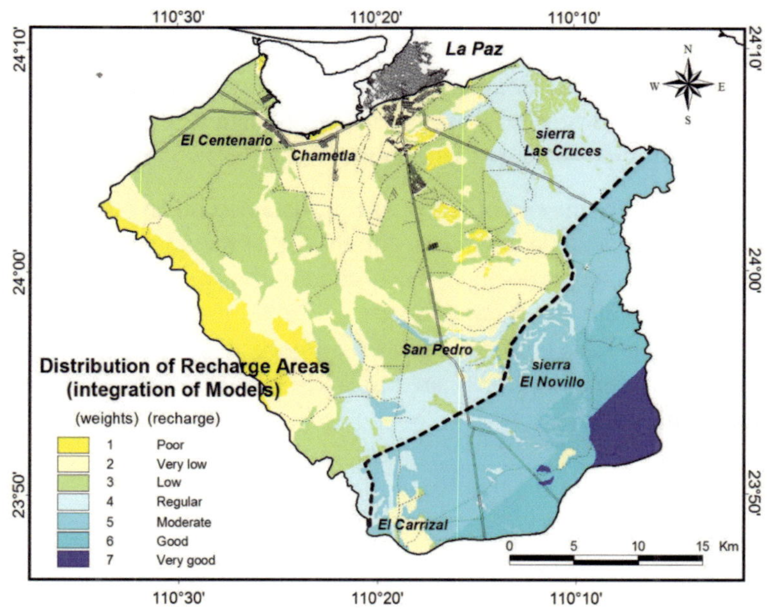

Figure 9. Model of rainfall recharge areas in the basin of La Paz. Dotted line delimits the best recharge areas (from moderate to very good).

4. Conclusions

The most important implications of the method of weights in the study of groundwater, enable to locate the rainfall recharge areas in the basin of La Paz. This includes spatial analysis and union of different layers, allowing to analyze and process each parameter in a simply and rapid way, by means of a geographical information system.

The use of the method of weights can handle the tendency or results based on the weight assigned to each class. For this study, the model of geology was given a higher weight to the fractured granite than to alluvium, due to authors criteria.

The final results of the processes involved with this method can be used, among other things, in watershed management, environmental management plans, urban development programs, and municipal strategic planning.

Due to the characteristics of the obtained models with its different classes, in relation to their contribution to recharge, it is observed that areas that best contribute to rainfall recharge are located on medium and high elevation zones within the basin. Poor, low and very low recharge areas, are distributed at the Center and Northwest of the basin, while regular recharge areas are distributed to the South and Northeast.

The areas classified from moderate to very good recharge, must be considered the best rainfall recharge areas of the basin of La Paz.

The best rainfall collecting areas, and rainfall recharge areas in the basin of La Paz, are located on the mountains and hills to the East and Southeast of the basin, around sierra El Novillo and Las Cruces, and the Northern portion of the valley of El Carrizal. Here, rainwater percolates underground within weathered igneous and metamorphic rocks, and through cracks and fractures. Then water is transported slowly down to the plains by groundwater flow. This process allows for recharge of the aquifer of La Paz, which is located in the valley.

At present, urban growth of the city of La Paz is not sustained on an ecological management program, neither on a sustainable urban development, that include ecological and economical alternatives for the preservation and conservation of the aquifer. In this sense, projected population growth of the city of La Paz towards the SSE of the basin, does not consider the natural rainfall collecting areas and recharge areas. Hence, it is very important and necessary to protect these areas, for being considered in land-use planning and urban development for the municipality of La Paz.

Acknowledgements

This work was supported by the Consejo Nacional de Ciencia y Tecnología (CONACyT), CONACyT-CIENCIA BASICA Fund, Project 134460 "Determinación y construcción de indicadores de la huella hídrica y desertificación como consecuencia de la sobreexplotación agropecuaria y del cambio climático en cuencas de zonas áridas", and by CIBNOR S.C.

Author details

Arturo Cruz-Falcón, Enrique Troyo-Diéguez, Héctor Fraga-Palomino and Juan Vega-Mayagoitia

*Address all correspondence to: afalcon04@cibnor.mx

Program of Agriculture in Arid Zones; Water, Soil and Weather, Centro de Investigaciones Biológicas del Noroeste (CIBNOR), La Paz Baja California Sur, México

References

[1] Álvarez, A. A. D., Rojas, S. H., & Prieto, M. J. J. (1997). Geología de la Bahía de La Paz y Areas Adyacentes. In: Urbán R. J. y R. M. Ramírez (editores), 1997. La Bahía de La Paz, Investigación y Conservación. UABCS-CICIMAR-SCRIPPS. , 13-29.

[2] Aparicio-Mijares F.J,(1992). Fundamentos de Hidrología de Superficie. Editorial Limusa, S.A. de C.V. Grupo Noriega Editores. México, D.F. 302 pp.

[3] Bridget, R. S., Healy, R. W., & Cook, P. G. (2002). Choosingappropiatetechniquesforquantifyinggroundwaterrecharge. HydrologeologyJournal , 10, 18-39.

[4] CIGSA (Consultores en Ingeniería Geofísica) S.A.de C.Estudio de Caracterización y Modelación de la intrusión Marina en el acuífero de La Paz B.C.S. Contrato CNA, GAS-013-PR01. 284 pp., 2001

[5] CNA (Comisión Nacional del Agua),(1997). Censo de Captaciones de Aguas Subterráneas y Colección de datos Geohidrológicos en la zona La Paz-El Carrizal, B.C.S. Informe final, Contrato GAS-PR097 para ADI Construcciones S.A. de C.V. Subdirección General Técnica, Gerencia de Aguas Subterráneas. 173 pp., 026.

[6] CNA (Comisión Nacional del Agua),(2005). Estudio para Actualizar la Disponibilidad Media Anual de las Aguas Nacionales Superficiales en las 85 (ochenta y cinco) Subregiones Hidrológicas de las 7 (siete) Regiones Hidrológicas 1,2,3,4,5,6 y 7 de la Península de Baja California, Mediante la Aplicación de la NOM-CNA-2000., 011.

[7] CONABIO (Comisión Nacional para el Conocimiento y Uso de la Biodiversidad), (2004). Capa vectorial de Geomorfología del Estado de Baja California Sur, escala 1:1,000,000.

[8] Cruz-Falcón, A., Vázquez-González, R., Ramírez-Hernández, J., Nava-Sánchez, E. H., Troyo-Diéguez, E., Rivera-Rosas, J., & Vega-Mayagoitia, y. J. E. (2011). Precipitación y Recarga en la Cuenca de La Paz, BCS, México. Universidad y Ciencia, , 27(3), 251-263.

[9] Davis, S. N., & De Wiest, R. (1971). Hidrogeología. Ed. Ariel, Barcelona, España. 563 pp.

[10] INEGI (Instituto Nacional de Estadística Geografía e Informática),(1983). Cartas Geológicas F12-3-5-6 y G12-10-11, escala 1:250,000.

[11] INEGI (Instituto Nacional de Estadística Geografía e Informática),(1998-2003). Cartas de Altimetría y Escurrimientos Superficiales, del conjunto de datos vectoriales. Cartas F12B11,12,13,14,24,33 y 34; G12D61,71,81,82,83 y 84, escala 1:50,000., 1998-2003.

[12] INEGI (Instituto Nacional de Estadística Geografía e Informática),(2000). Capas digitales F12-3-5-6 y G12-10-11 de Edafologíay Uso de Suelo y Vegetación del Estado de Baja California Sur, escala 1:250,000.

[13] IPN-CICIMAR/CIBNOR/UABCS (Instituto Politécnico Nacional- Centro Interdiciplinario de Ciencias Marinas / Centro de Investigaciones Biológicas del Noroeste / Universidad Autónoma de Baja California Sur),(2002). Ordenamiento Ecológico Bahía de La Paz, B.C.S. Informe preliminar. Abril, 2002.466 pp.

[14] Katpatal, Y. B., Dube, Y. A., & 200, . (2004). Geospatial Data Integration for Groundwater Recharge Estimation in Hard Rock Terrains. Department of Civil Engineering, Visvesvaraya National Institute of Technology. Maharashtra State, India. 10 pp.

[15] Kommadath, A. (2000). Estimation of Natural GroundwaterRecharge. Groundwater and Hydrogeology. http://ces.iisc.ernet.inSection 7, Paper 5. 7 pp.

[16] Peña-Haro, S., & Arcos-Hernández, D. (2004). Estimación inicial de la recarga vertical para su introducción a modelos de simulación de flujo con la ayuda de sistemas de información geográfica. Proyectos, Estudios y Sistemas, S.A. de C.V. Seminario Sistematización y automatización como herramienta para la gestión del agua, Expo-agua 2004. Comisión Estatal del Agua, Guanajuato, México. 9 pp.

[17] SARH-CNA,(1991). Sinopsis geohidrológica del Estado de Baja California Sur.Secretaría de Agricultura y Recursos Hidráulicos (SARH)-Comisión Nacional del Agua (CNA). Contrato No. AC-SH-Subdirección General de Administración del Agua, Gerencia de Aguas Subterráneas. 85 pp., 88-06.

[18] SARH-UNAM-UABCS,(1986). Estudio Geohidrológico complementario de las cuencas La Paz-El Carrizal, para proporcionar agua en bloque a la ciudad de La Paz, Baja California Sur. 334 pp.

[19] SEMARNAT,(2001). Presentación del concepto de cuencas ambientales en la SEMARNAT. Dr. Gustavo manuel Cruz Bello. Dirección General de Protección Ambiental e Integración Regional y Sectorial. México D.F. , 6-11.

[20] Sophocleous, Marios. A. (1991). Combining the Soilwater Balance and Water-Level Fluctuations Methods to Stimate Natural Ground water Recharge: Practical Aspects. Journal of Hydrology , 124, 229-241.

Chapter 8

Water Allocation Systems

Ralph A. Wurbs

Additional information is available at the end of the chapter

http://dx.doi.org/10.5772/54438

1. Introduction

Water resources are shared by many people who use the water for a variety of purposes. Water allocation systems serve to equitably apportion water resources among users; protect existing water users from having their supplies diminished by new users; govern the sharing of limited water during droughts when supplies are inadequate to meet all needs; and facilitate efficient water use. Effective water allocation becomes particularly important as demands exceed reliable supplies. As water demands increase with population and economic growth, water allocation systems must be expanded and refined.

The institutional framework for water resources development and management involves a hierarchy of water allocation systems. Water resources are allocated between nations by treaties and other agreements. In the United States (U.S.), water is allocated between states through interstate compacts. Water is shared by regional water authorities, municipal utility districts, cities, private companies, irrigation districts, farmers, and individual domestic water users through water rights systems. Water supply entities service their customers in accordance with contracts and other commitments. Federal water development agencies provide reservoir storage capacity for nonfederal sponsors.

This chapter begins with a general overview of institutional systems for allocating the water flowing and/or stored in rivers, lakes, and aquifers to diverse types of water use and numerous water users. The chapter then focuses on the Texas experience in implementing water allocation systems, including both accomplishments and issues still remaining to be resolved. The state of Texas in the United States serves as a case study that illustrates concepts and strategies that are relevant throughout the world.

With a population of 26 million people and land area of 696,000 km^2, Texas is a large state with diverse geography, economy, climate, hydrology, and water management practices. Texas has a rich heritage of implementing water allocation strategies as a central thrust of its water resources planning, development, and management. The Rio Grande is shared with Mexico, and several major river basins are shared with neighboring states in the U.S. Thousands of government agencies, cities, private companies, and citizens hold rights to use the waters of the state. Water allocation involves multiple overlapping institutional mechanisms. The state has progressed significantly in recent years in improving its water allocation systems. Further refinements continue to be a priority policy emphasis.

The Water Rights Analysis Package (WRAP) modeling system simulates the water allocation systems described in this chapter. WRAP is routinely applied in Texas in regional and statewide planning studies, administration of the water rights permit system, and other water management activities. WRAP is generalized for application anyplace in the world.

2. Water allocation mechanisms

People in various nations, regions, and local communities have developed their own sets of institutions and practices governing the sharing of water. These water allocation systems have evolved historically and continue to change. Hierarchies of water allocation systems in the U.S. and many other countries generally have the following components or features.

- The waters of international rivers and aquifers are allocated between nations based on international law, customs, treaties, and agreements.
- In the U.S., waters of interstate river basins are allocated between states based on compacts negotiated by the states and approved by the federal government.
- Certain rights are reserved for military installations and other government owned lands and facilities.
- A legally established priority system based generally on variations of the riparian or prior appropriation doctrines guides the allocation of the surface water flowing in streams and stored in reservoirs.
- A separate system of laws and customs guides the allocation of water resources in ground water aquifers.
- An administrative system that grants, limits, and modifies water rights and enforces the allocation of water resources may or may not include formal issuance of written permits to water right holders.
- Water users and water management entities implement various contracts and other formal agreements.
- Sharing of water resources may be governed by cultural traditions and informal agreements that evolve historically over many years.

Water allocation mechanisms typically vary greatly between ground-water and surface-water. From a water law perspective, ground and surface water are usually treated as separate resources. The extent to which the important hydrologic and water management interconnections are recognized varies between geographical regions.

The institutional mechanisms of water allocation are typically viewed from policy, legal, economic, and social perspectives. However, hydrologic science and engineering are also important aspects of developing and maintaining water allocation systems [1, 2, 3].

2.1. Allocation of the waters of international rivers and aquifers

Principles and rules of international water law are found in treaties, international custom, general principals of law, and writings of international institutions [4, 5, 6]. Two hundred and sixty-one international river basins, each encompassing portions of two or more nations, cover about 45 percent of the world's land area excluding Antarctica [7]. Little progress has been made in developing effective water allocation systems in many of these international river basins. Water allocation is even more difficult for groundwater aquifers shared by two or more countries.

Effective joint multiple-nation water management will be a major determinant in achieving stability, peace, and prosperity in many regions of the world in the 21st century [8, 9, 10, 11, 12]. Examples of the many regions with dramatic potential for either cooperation or conflict include the following. The Jordan River shared by Israel, Jordan, Syria, and the Palestinians is a small stream with remarkably great historical and political importance. Shared groundwater is also an important issue this region. The Euphrates and Tigris Rivers flow from Turkey through Syria, Iraq, and Iran. Most of the flow of the Euphrates and Tigris Rivers originate in their upper watersheds in Turkey and is controlled by a system of major dams in Turkey. The Ganges and Brahmaputra River Basins in Nepal, China, India, Bhutan, and Bangladesh, with a history of centuries of water conflicts, contained an estimated 400 million people in 2000 living at an impoverished standard of living. In the Southern African Region encompassing Angola, Botswana, Lesotho, Malawi, Mozambique, Namibia, South Africa, Swaziland, Tanzania, Zambia, and Zimbabwe, every major river is shared by two or more nations. Population growth and economic development are resulting in intensified demands on limited water resources with a long history of controversy.

2.2. Water rights in the United States

A water right is the legal privilege to store, regulate, and/or divert water for beneficial use. Water law is the creation, allocation, and administration of water rights. Institutions and customs for executing water rights play key roles in water management in various regions throughout the world. The following discussion is from a U.S. perspective but has broader applicability in other countries as well.

Books on water rights/law range from a concise book entitled *Water Law in a Nutshell* [13] to a massive eight volume collection [14] first published in 1967 and again as a revised updated edition in 1991 with the individual volumes continuing to be periodically updated and expanded at different times. The American Society of Civil Engineers has published guidelines for developing water right rules and regulations [15, 16, 17]. Other books focus on water rights in particular regions of the world other than the U.S. [18, 19, 20, 21].

Each state in the U.S. has developed its own set of laws, institutions, and practices governing water rights. These systems have evolved historically and continue to change. States in the western and eastern halves of the U.S. have generally adopted different approaches to water rights due largely to the western states having much drier climates. Water allocation and accounting systems tend to be more rigorous in regions where demands approach or exceed supplies. The experience of the state of Texas discussed in this chapter illustrates key aspects of developing and administering water right systems. Regions of Texas are representative of both western and eastern states.

Surface water in streams and lakes is viewed in the U.S. as a renewable resource owned by the state and used by the public. Groundwater is typically viewed as the property of the owners of the land overlying the aquifers. Water rights for groundwater aquifers are very different than for surface water streams and lakes. Although water rights are established primarily at the state level, federal laws govern water rights for military installations, other federal lands such as national parks, and Indian reservations.

For river basins encompassing portions of multiple states, water is allocated between states based on interstate compacts developed by the states, approved by the U.S. Congress, and administered by interstate compact commissions. Disputes may arise in implementing interstate compacts, particularly during droughts. Disputes are settled by the U.S. Supreme Court if compact commissions cannot work out disagreements between the states through negotiation. A major objective of compact commissions is to avoid lawsuits, but disputes in the western states in particular often reach the litigation stage. Various states have experienced lengthy and costly water allocation disputes settled through the U.S. Supreme Court [22]. Shared management of interstate groundwater aquifers and associated development of water allocation mechanisms have not progressed to nearly the extent as surface water.

2.2.1. Surface water rights

Legal rights to the use of stream flow are generally based on two alternative doctrines, riparian and prior appropriation. The basic concept of the riparian doctrine is that water rights are incidental to the ownership of land adjacent to a stream. The prior appropriation doctrine is based on the concept of protecting senior water users from having their supplies diminished by newcomers developing water supplies later in time. In a prior appropriation system, water rights are not inherent in land ownership, and priorities are established based on dates that water is appropriated.

The doctrine of riparian rights common in the eastern U.S. is based on English common law. Under the strictest interpretation of the riparian doctrine, the owner of riparian land adjacent to a stream is entitled to receive the full natural flow of the stream without change in quantity or quality. Since a strict interpretation imposes impractical constraints on water use, the riparian doctrine is normally interpreted to allow riparian land owners to divert reasonable amounts of stream flow for beneficial purposes. Varying definitions of *reasonable amounts* complicate water allocation.

The doctrine of prior appropriation is associated with settling the American West. As settlers moved from the eastern states to the West in the 1800's, farmers and ranchers claimed land, and miners claimed gold and other minerals. Likewise, water was appropriated by the first to arrive and claim the resources for beneficial use. In developing their farms and communities, people needed protection from having their water supplies diminished with later population growth and economic development.

Most of the western states have established permit systems in which a state agency issues permits to water right holders specifying amounts and conditions of water use. Riparian and/or appropriative rights may be incorporated into the original development of the permit system, with additional new permits being issued based on prior appropriation. With growing demands on limited water resources, permit systems will likely continue to be developed in the eastern states, which have more abundant stream flow, similar to those already in place in the drier western states.

Surface water rights in the eight driest western states (Nevada, Arizona, Utah, Idaho, Montana, Wyoming, Colorado, and New Mexico) are based purely on the prior appropriation doctrine. Alaska is also a prior appropriation state though somewhat different. Ten western states with hybrid systems merging riparian and appropriative rights into permit systems include California, Oregon, Washington, Texas, Oklahoma, Kansas, Nebraska, South Dakota, North Dakota, and Mississippi. Hawaii has a unique hybrid system. Water rights in 30 eastern states are based primarily on the riparian doctrine [23].

2.2.2. Groundwater rights

The rights and obligations for groundwater use are generally tied to two legal principles: property ownership and shared ownership of a common public resource. A variety of state approaches to groundwater rights has evolved from these concepts. State groundwater law is based on mixtures of the following doctrines.

Absolute Ownership Doctrine: Landowners own the groundwater under their land and may drill wells and pump as much water as they wish. Texas and several other states have historically adhered to this doctrine but are slowly changing.

Reasonable Use Doctrine: Landowners own groundwater, but their pumping is limited to reasonable use which has been defined in a variety of ways. This doctrine is common in the eastern states.

Correlative Rights Doctrine: In times of shortage, groundwater is shared by overlying landowners in proportion to the amount of land they own. This extension of the reasonable use rule is primarily associated with California.

Prior Appropriation Doctrine: Groundwater is allocated similarly to surface water with priorities assigned based on the dates that users first appropriate the water for beneficial use. This doctrine is common in the western states.

Permit Systems: Systems in which state agencies issue permits specifying the amounts and conditions of water use have been adopted in 33 of the 50 states. The other doctrines may be reflected in the water rights documented by the permits. Texas is among the 17 states that have no state-wide groundwater permit program.

Some states divide groundwater into categories with different water right rules applied to each classification. Percolating groundwater may be legally differentiated from underground streams with definable flow paths. Underground streams are sometimes treated as being similar to surface streams.

The issue of impacts of groundwater pumping on surface stream flow has been addressed to varying extents in different states. In some states, groundwater is classified as either tributary or nontributary. Tributary groundwater hydrologically contributes to surface stream flow. Nontributary groundwater does not. Water right rules and management strategies for tributary groundwater are based on protecting surface water rights.

3. Water allocation systems in Texas

Texas is a large state located in the south-central U.S. that is representative of both the drier western and wetter eastern states from various perspectives including climate. Mean annual precipitation varies from 20 cm at the city of El Paso in arid west Texas to over 140 cm in the humid eastern extreme of the state. Texas actually provides two case studies since water allocation in the Lower Rio Grande Valley has distinct differences from the remainder of the state. Water rights are a major consideration in river basin management statewide. Allocation of ground water is very different than allocation of surface water.

Water resources development and management in Texas is governed largely by the need to be prepared for extended droughts. The hydrologically most severe drought of record began gradually in 1951 and ended in 1957 with one of the largest floods on record. Droughts in the 1910's and 1930's were also extended multiple-year dry periods over large areas of Texas and neighboring states. The drought that occurred during 1995-1996 was much more economically costly than the earlier droughts due to the population and economic growth that had occurred. In terms of annual precipitation, for more than half of the land area of Texas, 2011 was the driest calendar year since the beginning of official observed precipitation records in 1895. The remainder of the state was also very dry during 2011. Severe drought conditions are continuing during 2012 throughout Texas and other states in the U.S.

3.1. Water resources planning, development, and management

The Texas Water Development Board (TWDB) in partnership with 16 regional planning groups outline future water resources needs and challenges in the 2012 State Water Plan, which is presented in a multiple volume report entitled *Water for Texas 2012* [24]. Water demands are projected to increase about 22 percent between 2010 and 2060. Available water supplies with existing infrastructure and current institutional arrangements will decrease about 10 percent during this period due to reservoir sedimentation and depletion of groundwater aquifers. Environmental flow requirements and ecosystem preservation are major concerns as well as meeting municipal, industrial, and agricultural water needs. The regional planning groups have identified several hundred water supply projects with an estimated cost of over $50 billion to address intensifying water needs. The TWDB predicts that annual losses from not meeting water supply needs could result in a reduction in income of approximately $12 billion annually if current drought conditions approach the drought of record, and as much as $116 billion annually by 2060 [24].

The map of Texas in Figure 1 shows the larger rivers and cities of the state. The state encompasses a land area of 696,000 km^2 divided into 15 major river basins and eight coastal basins located along the Gulf of Mexico between the lower reaches of the major river basins. The 1990 population of 17.0 million people increased to 25.7 million in 2011 and is projected by the Texas Water Development Board [24] to increase to 46.3 million by 2060. Fifty-eight percent of Texans live in the state's three largest metropolitan areas: the Dallas/Fort Worth Metroplex in the upper Trinity River Basin, greater Houston area in the San Jacinto River Basin, and city of San Antonio in the San Antonio River Basin. With a 2011 population of 6.5 million people, the Dallas/Fort Worth (DFW) Metroplex is the largest metropolitan area in the southern U.S. and the fastest growing metropolitan area in the U.S. With a 2011 population of 6.1 million, the city Houston and adjacent smaller cities is also one of the largest and fastest growing metropolitan areas in the U.S. San Antonio is the third largest city in Texas with a population of 2.2 million people. Conversely, several of the major river basins of the state have extremely low population densities. Water management practices are very diverse throughout the state.

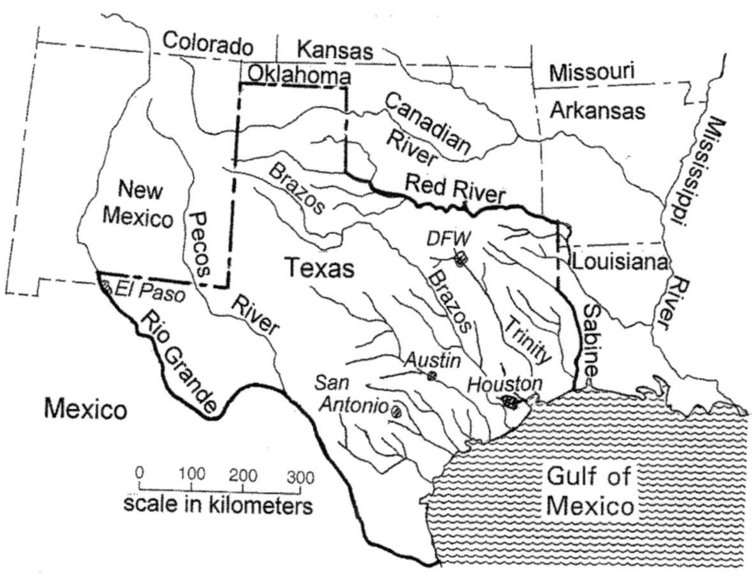

Figure 1. Map of Texas showing major rivers, largest cities, and neighboring states

Total diversions of 22 billion m^3/year from streams, reservoirs, and aquifers statewide in 2010 was for agricultural irrigation (56%), municipal use (27%), industrial use (15%), and livestock (2%). Currently, about half of the water supplied is from groundwater aquifers, and the other half is from surface streams and reservoirs. However, problems caused by decades of groundwater pumping rates significantly exceeding recharge rates in various regions of the state are resulting in major shifts toward greater reliance on surface water.

The naturalized flow volumes during each of the 864 months of 1940-2011 at a gauging station on the Brazos River near Houston are plotted in Fig. 2 to illustrate the tremendous variability that characterizes stream flows throughout Texas. A highly variable resource is allocated to numerous water users. Long-term mean flows may be relatively large, but most of the flow occurs during infrequent flood events. The water management community must deal with droughts with durations ranging from several months to several years. Dams with large storage volumes are essential in order to develop dependable water supplies. The river flows plotted in Fig. 2 are generated by the modeling system described later in this chapter by adjusting gauged flows to represent natural conditions without human water resources development and use.

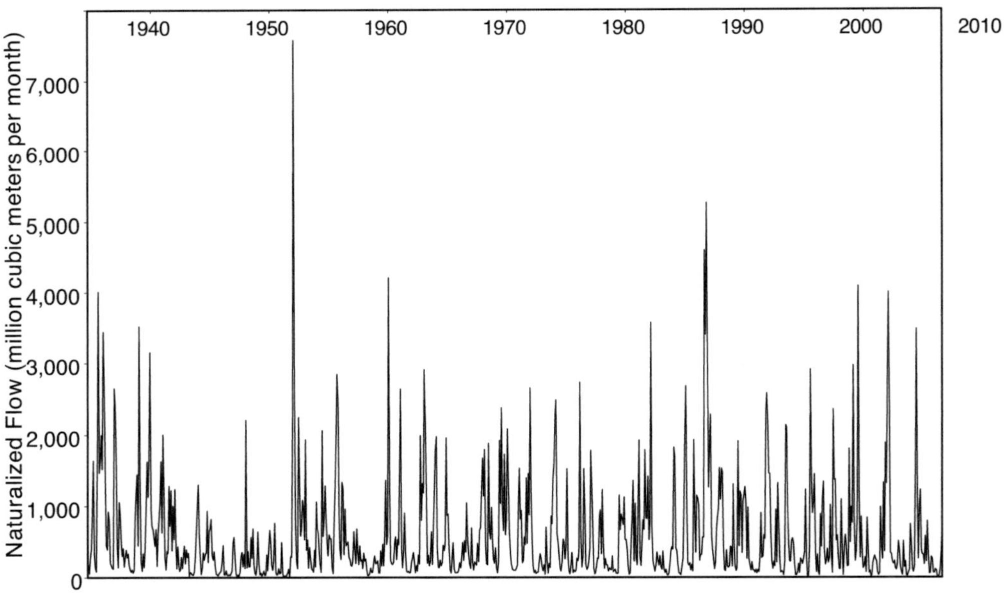

Figure 2. Monthly naturalized flows of the Brazos River illustrating variability

Conservation and flood control storage capacities totaling 50 and 23 billion m^3 are provided by 211 major reservoirs with individual capacities of 6.2 million m^3 or greater. Although there are several thousand smaller reservoirs, the 211 major reservoirs account for over 95% of the storage capacity in the state. Texoma on the Red River and International Amistad on the Rio Grande are the largest reservoirs with capacities of 6.6 and 6.3 billion m^3. The 22 hydroelectric plants in the state are components of electric utility systems dominated by thermal plants. Reservoir releases through hydropower turbines are almost always incidental to downstream water supply needs.

Federal involvement in developing the state's water resources includes 32 U.S. Army Corps of Engineers (USACE) reservoirs in eight river basins and two International Boundary and Water Commission (IBWC) reservoirs on the Rio Grande that account for about 40% and 90% of the conservation and flood control, respectively, storage capacity of the 211 major reservoirs. The U.S. Bureau of Reclamation constructed three reservoirs that are now owned by local entities. River authorities and cities have contracted for the water supply storage capacity of the USACE reservoirs. All costs allocated to water supply are reimbursed by the nonfederal sponsors, who also hold the water right permits. The USACE is responsible for flood control operations.

Nineteen river authorities are responsible for development and management of the water resources of all or portions of major river basins. River authorities construct and operate their own reservoir projects and contract for storage capacity in federal reservoirs. Numerous municipal utility districts and irrigation districts also play key roles in supplying water from rivers and reservoirs. Groundwater conservation districts facilitate groundwater management.

Some type of water rights system has been administered statewide by a centralized agency since 1913, but that agency has changed over time. The Board of Water Engineers was established in 1913; reorganized as the Texas Water Commission (TWC) in 1962; and renamed the Texas Water Rights Commission in 1965 with non-water rights functions being transferred to the Texas Water Development Board (TWDB), which had been previously created in 1957. In 1977, the Texas Department of Water Resources (TDWR) was created by combining the Water Rights Commission, TWDB, and Water Quality Board. In 1985, the TDWR was dissolved, and the TWC and TWDB became separate agencies. The Texas Natural Resource Conservation Commission (TNRCC) was created in 1993 by merging the TWC and Texas Air Quality Board. The TNRCC was renamed the Texas Commission on Environmental Quality (TCEQ) in September 2002. The TCEQ is now responsible for administering the many regulatory programs of the state including the water rights system.

The TCEQ consists of three full-time commissioners appointed by the governor and a professional and administrative staff of over 3,000 employees. Water rights are one of many regulatory responsibilities of the TCEQ. The TCEQ and TWDB interact closely in many of their activities. The TWDB is responsible for developing and updating the State Water Plan and administering an array of financial assistance programs. The TWDB has a governing board of six members appointed by the governor and a staff of about 800 employees.

3.2. Legislatively mandated water management programs

The state-level water allocation systems described in this chapter were created under laws enacted by the Texas Legislature. As discussed later, the statewide water rights permit system for all of Texas except the Rio Grande was implemented pursuant to the Water Rights

Adjudication Act of 1967. The water rights adjudication process mandated by the 1967 Act was completed shortly before a severe statewide drought in 1995-1996 which motivated enactment of the 1997 Senate Bill 1.

The Brown-Lewis Water Management Plan enacted by the Texas Legislature as its 1997 Senate Bill 1 is considered to be a milestone in the history of water resources management in Texas. The designation *Senate Bill 1* is traditionally applied each legislative session to highlight legislation of the upmost importance. The 1997 Senate Bill 1 and subsequent amending legislation authorized the Water Availability Modeling (WAM) System described later in this chapter, a regional and statewide water resources planning process noted below, and other related water management activities.

The 1997 Senate Bill 1 established a program for developing regional water plans that are integrated into a statewide planning process administered by the Texas Water Development Board (TWDB). Committees of local water interests have been established to prepare plans for the orderly development, management, and conservation of the water resources of each of 16 regions. The TWDB provides funding, administrative, and technical support to the regional committees. Consulting firms perform much of the technical work. The 1997 Senate Bill 1 mandated that initial regional plans be completed and incorporated into a statewide plan by 2002. Continuing planning is organized based on updated plans being reported at cycles of not to exceed five years. The 2002, 2007, and 2012 state water plans consists of 16 regional reports and a statewide report which are publically available at the TWDB website.

In evaluating water right permit applications, the TCEQ requires that proposed actions must be consistent with pertinent regional plans. The WAM System is routinely applied in both regional and statewide planning and administration of the water right permit system.

The Texas Instream Flow Program [25] established by Senate Bill 2 (SB-2) enacted by the Texas Legislature in 2001 and expanded by Senate Bill 3 (SB-3) in 2007 is administered jointly by the TCEQ, TWDB, and Texas Parks and Wildlife Department (TPWD). The purpose of the SB-2 program is to perform scientific studies to determine flow conditions necessary for supporting a sound ecological environment in the river basins of Texas. SB-3 mandated a new regulatory approach for protecting environmental flows through a local stakeholder process culminating in rules to be administered through the TCEQ. The objectives of the 2001 SB-2 and 2007 SB-3 are being accomplished largely through the ongoing work of Bay and Basin Expert Science Teams (BBEST) and Bay and Basin Area Stakeholder Committees (BBASC) for the individual river basins, representing the scientific and water management/use communities, with technical support from consultants. The TCEQ, TWDB, and TPWD provides administration oversight and technical support.

Environmental instream flow requirements are determined by the BBEST teams for their assigned streams within a framework of subsistence flows, base flows, high flow pulses, and overbank flood events needed to support riverine ecosystems, wetlands, and freshwater inflows to bays and estuaries. The Stakeholder Committees consider the BBEST recommendations within the content of municipal, industrial, and agricultural water needs as well as environmental water needs. The TCEQ is responsible for final approval and incorporation of the recommendations of these groups and the public into the water rights permit system. As of 2012, the endeavors of the Texas water management community, under the leadership of the three-agency partnership and the BBEST and BBASC groups, to better incorporate environmental considerations into the water rights permit system is well underway and is expected to continue for many years into the future.

3.3. Groundwater management

Most of Texas is underlain by 9 major and 21 minor aquifers that currently supply about half of the water used in the state. Depleting ground water reserves due to excessive pumping is a major problem throughout the state and is forcing a shift from groundwater use to a greater reliance on surface water supplies. Groundwater rights in Texas have historically been based on the common law rule allowing land owners to pump as much water as they wish from under their land. However, increased state regulation of groundwater is evolving over time primarily through the establishment of groundwater conservation districts.

In 1949 and 1985, the Texas Legislature passed laws authorizing creation of groundwater conservation districts with local voter approval. The 1949 legislation allows local residents to petition the state to have a district created. The 1985 amendment authorizes the TCEQ to designate and initiate formation of districts for areas with critical problems. Local voters can still veto a proposed district, but if they do, state funding for water projects can be withheld. Twelve districts existed prior to 1985. As of 2012, 96 groundwater districts covering over half of the land area of the state are operational. Three additional districts are currently in the process of being confirmed by voters through local elections at the county level. A total of 174 of the 254 counties in Texas are within a groundwater conservation district.

The primary purposes of the districts are to encourage water conservation and to protect water quality. Most districts direct their efforts toward prevention of waste, water conservation education, recharge projects, and data collection. Some are now moving toward stricter regulation of groundwater use. The districts tread a narrow path between private ownership of groundwater and state responsibility to protect the water resource. Texans are reluctant to allow anyone to tell them how much water they can pump from under their own land. Governmental regulation of pumping has been driven by necessity as depleting aquifers resulted in major problems. The Harris-Galveston Coastal Subsidence District and Edwards Underground Water Authority have developed the strongest regulatory programs.

The Harris-Galveston Coastal Subsidence District was created in 1957 in response to severe subsidence in the vicinity of the cities of Houston and Galveston. Due to decades of overdrafting groundwater, the ground surface has been lowered over three meters in places in this low-lying, heavily urbanized coastal region. Houston and neighboring smaller cities continue to shift from ground water to surface water supplies. Groundwater pumping is regulated by the Subsidence District through a permit program.

The Edwards Aquifer Authority was created in 1993 largely due to a federal court ruling related to protection of endangered species under the Endangered Species Act. The Edwards is a limestone aquifer shared by San Antonio, several smaller cities, and extensive irrigated farming interests. San Antonio is the largest city in the U.S. that relies solely on groundwater for its water supply. Springs fed by the Edwards Aquifer maintain the flow of several rivers and support ecosystems that include several endangered species. The Edwards Aquifer Authority administers a water rights permitting system to facilitate aquifer management. Pumping limits are reduced whenever water levels in selected wells fall below specified criteria.

All of the groundwater conservation districts are required to develop, implement, and periodically update management plans for the effective management of their groundwater resources. These plans are subject to approval by the TWDB and are publically accessible through the TWDB website.

Although interstate and international aquifers are important, allocation agreements do not exist. The U.S. and Mexico have different legal reasons for their lack of management of ground water [26]. The Mexican federal government has the authority to enforce comprehensive management and regulation of ground water but has not chosen to do so. In the U.S., the legal framework is inadequate and chaotic. Another constraint to joint U.S. and Mexico management of ground water is the lack of clarity of international ground water law [26].

3.4. Allocation of the waters of the Rio Grande

The Rio Grande (Spanish for Large River) is unique relative to the other river basins of Texas from several perspectives. It is shared by two nations. The Lower Rio Grande Valley accounts for the majority of the surface water irrigation in Texas. The intensive agricultural production of the region depends almost exclusively on the Rio Grande with little use of groundwater. Other major irrigation regions of the state rely primarily on groundwater. The water rights system for the Lower Rio Grande was developed separately and has distinct differences from the remainder of the state, particularly in regard to the priority system and water master operations. Fort Quitman shown in Fig. 3, located 140 km downstream of the city of El Paso, is used as the divide between the lower and upper portions of the basin in both the international and state water allocation systems.

The Rio Grande Basin is shared by Mexico and three states in the U.S. Water allocation is governed by two international treaties and two interstate compacts. Allocation of the Texas share of the waters of the Rio Grande to irrigators, cities, and other users is based on a water rights permit system governed by state law. The Rio Grande Compact approved by the legislatures of Colorado, New Mexico, and Texas in 1939 allocates the uncommitted waters of the Rio Grande above Fort Quitman. The Pecos River Compact adopted in 1948 allocates the waters of that tributary between Texas and New Mexico.

3.4.1. Mexico-United States treaties

A 1906 treaty between the U.S. and Mexico provides for delivery of 74 million m^3/yr of Rio Grande water to Mexico in the El Paso-Juarez Valley above Fort Quitman. Elephant Butte Reservoir in New Mexico, operated by the Bureau of Reclamation, and the American and

http://dx.doi.org/10.5772/54438

International diversion dams near El Paso, operated by the IBWC are used to implement the water allocation provisions of the treaty. The treaty further provides that if water is unavailable, the amount allocated to Mexico shall be diminished in the same proportion as water delivered to irrigate lands in the United States. This provision has been invoked in about a third of the years since 1951.

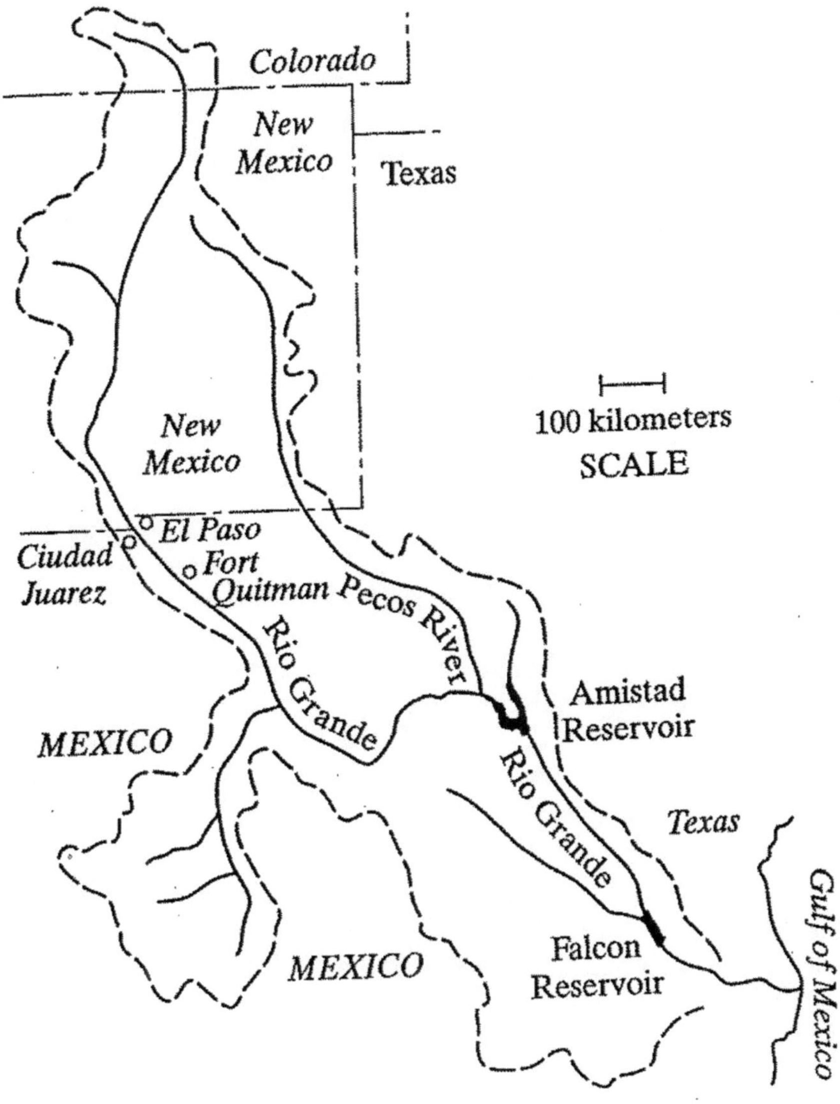

Figure 3. Rio Grande Basin

The Water Treaty of 1944 expanded the International Boundary Commission to the International Boundary and Water Commission (IBWC), provided for the distribution of waters of the Rio Grande from Fort Quitman to the Gulf of Mexico between the two nations, and authorized construction of International Amistad and Falcon Reservoirs.

The 1944 treaty administered by the IBWC also includes provisions for allocation of the waters of the Colorado River. The Colorado River drains 629,000 km^2 in seven western states and flows into the Gulf of California in Mexico. Although both are included in the same 1944 treaty, water allocation for the Colorado River and Rio Grande are very different. The Rio Grande serves as a boundary between the two countries. The Colorado River flows from the U.S. to Mexico with most of its watershed area being located in the U.S.

The IBWC is composed of a Mexico Section with commissioner and technical and administrative staff located in Ciudad Juarez and a U.S. Section with offices across the river in El Paso. Amistad and Falcon Reservoirs are operated by the IBWC primarily for flood control and water supply for the Lower Rio Grande Valley. Hydroelectric power generation uses releases for downstream water supply. The U.S. share of the conservation storage is used to meet demands in the lower basin in accordance with the state water rights system.

Conservation and flood control capacities are 4.17 billion m^3 and 2.15 billion m^3 in Amistad Reservoir and 3.29 and 0.63 billion m^3 Falcon Reservoir. In accordance with the 1944 treaty, the U.S. has 56.2% and 58.6% of the conservation storage capacity of Amistad and Falcon, respectively, with Mexico owning the remainder. The IBWC operates Anzaldus and Retamal Dams on the lower Rio Grande to facilitate diversions. The travel time for releases from Falcon Reservoir to reach the most downstream diversion locations is about a week.

The IBWC maintains a continuous accounting of the volume of stored water owned by each of the two countries. Stream flows into Falcon and Amistad Reservoirs are allocated between the two countries. Flows on a number of major tributaries named in the treaty are gauged and allocated as specified by the treaty. All other flows not otherwise allocated are divided equally between the two countries. Mexico receives all the flows from several specified Mexican tributaries and two-thirds of the flows from other specified tributaries. The U.S. receives all the flows from certain U.S. tributaries and one-third of the flows from the other specified tributaries. Computations are performed routinely to allocate the reservoir inflow and evaporation volumes, which are combined with recorded releases to determine the amount of water that each country has in storage.

The 1944 treaty gives the U.S. a right to one-third of the flow reaching the Rio Grande from six Mexican tributaries, provided that this third shall not be less, as an average amount in cycles of five consecutive years, than 431,721,000 m^3 annually. Other provisions relate to special conditions. A significant deficit accumulated over several years during the 1900's-2000's in meeting these requirements. Discussions addressing the stream flow deficit owed by Mexico to the U.S. included requests by the U.S. that operating policies for reservoirs in Mexico be modified to mitigate the deficit.

3.4.2. Allocation of the Texas share of the lower Rio Grande

The Texas share of the waters of the Rio Grande below Fort Quitman was allocated among numerous water rights holders in conjunction with a massive lawsuit, commonly called the

Lower Rio Grande Valley Water Case. The lawsuit was filed in 1956, the trial was held in 1964-1966, and the final judgment of the appellate court was filed in 1969. In 1971, the Texas Water Rights Commission adopted rules and regulations implementing the court decision. Assorted versions of riparian and appropriative rights were combined into a permit system. The litigants in the Rio Grande law suit included 42 water districts and 2,500 individuals. More than 90 lawyers appeared before the court. The expense and effort demonstrated the impracticality of a purely judicial determination of water rights for the entire state and led to enactment of the Water Rights Adjudication Act of 1967.

The lawsuit resulted in water rights being divided into three categories. Municipal rights have the highest priority. Irrigation rights are divided into Class A and Class B rights, with Class A rights receiving more storage in Falcon and Amistad Reservoirs storage accounts in the allocation procedure. Although this weighted priority system for irrigation rights has little significance during years of plentiful water, its effect in water-short years is to distribute the shortage among all users, with the greater shortages occurring on lands with Class B water rights.

Most of the U.S. share of the water regulated by Amistad and Falcon Reservoirs is used in the very productive agricultural region of Texas below Falcon Reservoir. Irrigation districts, individual farmers, and cities communicate their water needs to the TCEQ Rio Grande Water Master Office, which in turn schedules releases from Falcon and Amistad Reservoirs with the IBWC. The Water Master Office maintains an accounting of the amount of water used and the amount of water in reservoir storage allocated to each of about 1,600 water rights accounts.

The allocation rules administered by the TCEQ Water Master first provide a reserve of 278 million m^3 in Amistad and Falcon Reservoirs for domestic, municipal, and industrial uses. Next, available water is allocated to an operating reserve that provides for seepage and evaporation losses, adjustments required as the IBWC computations of Mexico-U.S allocations are updated periodically, and emergency requirements. The remaining water in storage is allocated among all the irrigation permit holders. The storage is basically allocated in proportion to annual diversion rights, except the Class A rights are multiplied by a factor of 1.7 to allow them a greater storage allocation than Class B rights. Other provisions include limiting each storage allotment to not exceed more than 1.41 times its authorized diversion right. If an irrigation right does not use water for two consecutive years, its storage account is reduced to zero.

3.5. Interstate river compacts

Texas participates in five interstate river compacts executed by the member states and approved by the U.S. Congress. The rivers and the dates the compacts became effective are Rio Grande, 1939, Pecos, 1948, Canadian, 1952, Sabine, 1954, and Red, 1980. The purposes of the compacts are to provide for equitable allocation of water between the states and to facilitate cooperative planning and management. Commissions with a representative from each member state administer the compacts. The commissioners have minimal staffs and rely on the state water agencies for technical support.

Texas filed a lawsuit against New Mexico in 1974, claiming that Texas did not receive its full share of water allocated by the 1948 Pecos River Compact due to over-use in New Mexico [27]. In 1988, based on the recommendations of a court appointed Special Master, the Supreme Court awarded Texas a cash settlement for past damages and established a Pecos River Master to administer future water allocations under the compact.

3.6. Statewide surface water rights permit system

All of Texas except the Lower Rio Grande below Fort Quitman has a consistent surface water rights permit system which is different than the system just described for the Lower Rio Grande. The TCEQ administers both systems. River authorities, municipal utility districts, cities, irrigation districts, farmers, companies, and citizens hold about 8,000 permits statewide. The water is owned by the state, with rights granted to organizations and people to use prescribed amounts for prescribed purposes under prescribed conditions.

3.6.1. Historical evolution of water rights

The riparian doctrine was introduced in Texas during the rule of Spain and later Mexico and after independence in 1836, in a somewhat different form by the Republic of Texas. In 1840, the state of Texas adopted the common law of England with another variation of riparian water rights. The extent to which riparian rights allow large irrigation developments or other large amounts of use depends upon the laws in effect when the land was originally transferred from public to private ownership. Riparian rights are different depending on whether the land can be traced to land grants from Spain, Mexico, Republic of Texas, or the state of Texas.

The prior appropriation doctrine was adopted by legislative acts in 1889 and 1895. After 1895, public lands which transferred into private ownership no longer carried riparian water rights. Land already privately owned kept its riparian rights. At first, appropriation simply involved water users filing sworn statements with county clerks describing their use. Since 1913, more strictly administered procedures have been followed based on a statewide appropriation system administered by a centralized state agency.

All appropriation statutes recognized existing riparian rights. Riparian landowners could also acquire appropriative water rights. In 1926, the courts divided stream flow into *ordinary normal flow* and *flood flow*. Riparian rights were limited to normal flow and therefore are not applicable to flood waters impounded by reservoirs. Although difficult to apply in actual practice, this distinction was the basis for correlating riparian and appropriative rights from 1926 until the riparian rights were merged into the appropriative system pursuant to the Water Rights Adjudication Act of 1967.

Thus, an unmanageable system evolved, with various conflicting rights and many rights being unrecorded. The 1951-1957 drought motivated the previously noted Lower Rio Grande Valley Water Case which clearly demonstrated the impracticality of a purely judicial adjudication of water rights statewide. Thus, the Water Rights Adjudication Act was enacted by the Texas legislature in 1967, with the stated purpose being to require a recording of all

http://dx.doi.org/10.5772/54438

claims for water rights, to limit the exercise of those claims to actual use, and to provide for the adjudication and administration of water rights. All riparian water rights were merged into the prior appropriation system, creating the present permit system applicable to all of Texas except the Lower Rio Grande. The water rights adjudication process required for transition to the permit system was initiated in 1968 and completed by the late 1980's.

3.6.2. Prior appropriation permit system

Water rights are granted by a state license, or permit, which allows the holder to divert a specified amount of water annually at a specific location, for a specific purpose, and to store water in reservoirs of specified capacity. Any organization or person may submit an application to the TCEQ for a new water right or to change an existing water right at any time. The TCEQ will approve the permit application if unappropriated water is available, a beneficial use of the water is contemplated, water conservation will be practiced, existing water rights are not impaired, and the water use is not detrimental to the public welfare. Proposed actions reflected in water right permit applications must be consistent with regional water plans.

The water authorized to be appropriated under the terms of the particular permit is not subject to further appropriation unless the permit is canceled. A permit may be canceled if water is not used during a 10-year period. Special term permits may also be issued allowing water use for specified periods of time. The Rio Grande and segments of other rivers are over-appropriated with no new rights for additional water use being granted. However, unappropriated water is still available for appropriation in other regions of the state.

A permit holder has no actual title of ownership of the water but only a right to use the water. However, a water right can be sold, leased, or transferred to another person. The Lower Rio Grande Valley has been the only region of Texas with an active water market historically. In 1993, the legislature established a statewide water bank to be administered by the TWDB. Although transfers can be accomplished independently of the water bank, the program was created to encourage and facilitate water marketing, transfer, and reallocation.

The Texas Water Code lists beneficial uses in order of priority as follows: (1) domestic and municipal, (2) industrial, (3) irrigation, (4) mining, (5) hydroelectric, (6) navigation, (7) recreation and pleasure, and (8) other beneficial uses. These priorities are invoked only when a conflict exists between water use applications. Under the prior appropriation system, after a permit is in effect, priorities are based on the date specified in the permits. During the 1968-1980's adjudication process, priority dates were established based on historical water use. Since then, priorities are based on the dates that the permits are issued. In times of emergency, cities may take water even if non-municipal users are adversely affected, regardless of priority dates.

4. Water Rights Analysis Package (WRAP) modeling system

The Water Rights Analysis Package (WRAP) computer modeling system is generalized for application to river/reservoir/use systems located anywhere in the world, with model-users developing input datasets for the particular river basin of concern. For applications in Texas, publicly available WRAP input datasets from the TCEQ Water Availability Modeling (WAM) System [28] are altered as appropriate to reflect proposed water management plans of interest, which could involve changes in water use or reservoir/river system operating practices, construction of new facilities, or other water management strategies.

WRAP simulates water resources development, management, regulation, and use in a river basin or multiple-basin region under a priority-based water allocation system. In WRAP terminology, a water right is a set of water use requirements, reservoir storage and conveyance facilities, hydropower projects, operating rules, and institutional arrangements for managing water resources. Stream flow and reservoir storage is allocated among users based on specified priorities, which can be defined in various ways. Simulation results are organized in optional formats including time sequences of many different variables, summaries, water budgets, frequency relationships, and various types of reliability indices.

The public domain WRAP software and documentation [29, 30, 31, 32, 33] may be downloaded free-of-charge from http://ceprofs.tamu.edu/rwurbs/wrap.htm. The WRAP website links to TCEQ WAM and TWRI websites that provide reports, datasets, and other related information. Wurbs provides a concise summary of WRAP [34] and comparison with other generalized modeling systems [35].

WRAP modeling capabilities that have been routinely applied in the Texas WAM System consist of using a hydrologic period-of-analysis of about 60 years and monthly time step to perform water availability and reliability analyses for municipal, industrial, and agricultural water supply, environmental instream flow, hydroelectric power generation, and reservoir storage requirements. Recently developed additional WRAP modeling capabilities include: short-term conditional reliability modeling [36]; daily time step modeling capabilities that include flow forecasting and routing and disaggregation of monthly flows to daily; simulation of flood control reservoir system operations [31]; and salinity simulation [37]. Further improvements to WRAP currently underway, as of 2012, are focused on better integrating environmental flow requirements into comprehensive water management.

The generalized WRAP modeling system was developed and continues to be expanded at Texas A&M University (TAMU). Early versions dating back to the 1980's were developed under the sponsorship of the Texas Water Resources Institute (TWRI) and U.S. Department of Interior. Efforts at TAMU to expand and improve WRAP from 1996 through the present have been sponsored primarily by the TCEQ. However, the TWDB, U.S. Army Corps of Engineers, Brazos River Authority, and other agencies have funded specific improvements to WRAP. Model development has been an evolutionary process with extensive interactions between professionals from the agencies and consulting firms applying the model to specific river basins and university researchers responsible for improving the modeling methodology and computer software.

http://dx.doi.org/10.5772/54438

4.1. Texas water availability modeling system

As previously noted, water management legislation known as Senate Bill 1 enacted by the Texas legislature in 1997 directed the water agencies to establish a regional planning process and a water availability modeling system. The Texas WAM System was implemented by the TCEQ, its partner agencies (TWDB and TPWD) and contractors (ten consulting engineering firms and two university research entities) during 1997-2002 pursuant to the 1997 Senate Bill 1. The WAM System has continuously to be improved and expanded since 2002 along with being routinely applied by the water management community [28].

The Water Availability Modeling (WAM) System consists of the generalized WRAP and input datasets for the 23 river basins of Texas. The WAM datasets include naturalized stream flows at a total of about 500 gauged sites, watershed parameters for distributing these flows to over 12,000 ungauged locations, 3,450 reservoirs, various other constructed infrastructure, operating plans that in many cases are quite complex, two international treaties, five interstate compacts, various contractual agreements, and water use requirements associated with about 8,000 water right permits reflecting two different water right systems.

Prior to creation of the WAM System, many water right permit holders incorrectly assumed that the amount of water specified in their permits would always be available to them. Senate Bill 1 required that the TCEQ notify all permit holders regarding the reliabilities associated with their permits. All water needs cannot be met during severe droughts.

The TCEQ requires that permit applicants, or their consultants, apply the WRAP/WAM system to assess supply reliabilities and flow and storage frequencies associated with their proposed actions and the impacts on all of the water users in the river basin. TCEQ staff applies the modeling system to evaluate the permit applications. The TWDB and regional planning groups apply the modeling system in their planning studies. The TCEQ requires that water right permit applications be consistent with regional plans. River authorities and other water suppliers apply the modeling system in operational planning studies.

4.2. Lessons learned in implementing the WRAP/WAM system

Developing and applying computer models have typically been viewed in terms of the engineering and scientific concepts and methods incorporated in the models. However, modeling has important institutional as well as technical dimensions. Lessons learned from development and application of the Texas WAM System demonstrate the importance of the following two institutional dimensions of river/reservoir system modeling.

1. Modeling water rights, contractual agreements, treaties, interstate compacts, and other complex institutional aspects of water resources development, management, allocation, and use may be a key consideration in developing and applying a modeling system.

2. Effective implementation of a modeling system may require a partnership effort of an entire water management community that includes water users, political officials, legislatures, environmental and other non-governmental special interest groups, government agencies, consulting firms, and university researchers.

The following general observations characterize the Texas experience in implementing a water availability modeling system.

- Droughts motivate political concern, improvements in water management, and development of computer modeling systems.
- Partnerships and consensus building are key aspects of water resources planning and management. Likewise, a water management community may work together to effectively implement a shared modeling system. Model development and application may be an institutional partnership effort.
- Administration of water allocation systems has become a central focus of river basin management. Regulatory and planning functions are integrally related. Shared modeling tools can facilitate integration of planning and regulatory functions.
- Modeling systems include computer programs, databases, organizations, people, and decision-making processes. Compilation and management of voluminous data is a central governing concern. A modeling system is constructed rather than just a model.
- Model development is a dynamic evolutionary process. As long as a computer simulation model such as WRAP continues to be applied, its development is never completed. Model development is a process of continual expanding and improving.
- Water availability modeling is essential for effective water management.

5. Water allocation issues

Allocating water resources that are highly variable both temporally and spatially among a myriad of water management entities and numerous water users within an institutional setting that has evolved historically over many years is necessarily complex. The following concerns highlighted by the Texas experience are illustrative of the numerous complexities in creating and administering water allocation systems.

For most of Texas, the water right permit system is administered without water master operations. Upon request, the TCEQ takes enforcement action to stop reported unauthorized water use in violation of water rights permits. However, water users are not closely monitored except during droughts or emergency conditions. This approach is similar to most western states. Several western states have water-master operations, but most states do not. The TCEQ during 2012-2013 is investigating the feasibility of expanding water master operations in Texas.

The TCEQ Lower Rio Grande Water Master Office maintains a precise accounting of water use, working closely with irrigators, cities, and the International Boundary and Water Commission. With completion of the adjudication process during the late 1980's, plans were developed to establish water-master operations in all of the major river basins of Texas. The South Texas Water-Master Program was created in the late 1980's with responsibilities for

http://dx.doi.org/10.5772/54438

the Guadalupe, Nueces, and San Antonio River Basins. However, water users are reluctant to have requirements imposed upon them for installing meters and monitoring and regulating diversions. Political pressures have prevented the establishment of water-master offices in other river basins. However, the Texas Legislature in 2012 directed the TCEQ to solicit public input and develop recommendations for establishing water master operations for other river basins.

Since stream flow, evaporation, reservoir sedimentation, water use, and other factors are highly variable, and the future is unknown, water availability must be viewed from a reliability, likelihood, or percent-of-time perspective. Tradeoffs occur between the amount of water to commit for beneficial use and the level of reliability that can be achieved. Beneficial use of water is based on assuring a high level of reliability. However, if water commitments are limited as required to assure an extremely high level of reliability, the amount of stream flow available for beneficial use is constrained, and a greater proportion of the water flows to the ocean or is lost through reservoir evaporation. The optimal level of reliability varies with type of water use. Water allocation decisions necessarily require qualitative judgment in determining acceptable levels of reliability both in terms of the reliability of the proposed new or increased water use and the impacts on the reliabilities of all of the existing water users.

Many of the existing water rights adjudicated pursuant to the 1967 Water Rights Adjudication Act and the Lower Rio Grande Court Case as well as some recently established rights have supply reliabilities that are much lower than the water management community considers desirable. However, the TCEQ has applied more stringent criteria in approving newer water right permits or modifications to permits. In evaluating water right permit applications for agricultural irrigation, the TCEQ now applies the criterion that at least 75 percent of the proposed demand should be supplied at least 75 percent of the time as determined by the WRAP/WAM System. The TCEQ criterion for municipal permit applications is that 100 permit of the demand should be supplied 100 permit of the time based on the premises reflected in the WRAP/WAM model including historical hydrology. However, these criteria may be modified depending upon backup water supply sources. For example, with depleting groundwater reserves, a transition to a surface water source may be worthwhile, even if the reliability of the surface water source is low, if groundwater can still be used as a backup supply.

Although Texas and other western states are viewed as adopting the prior appropriation doctrine, strictly speaking a pure prior appropriation system is not feasible and does not exist. Although the effect may be very small, developing additional supplies for new users always affects downstream supplies. Also, in drought situations, water supply shortages are shared, to some degree, by water users regardless of the relative seniority of their rights. Sharing of water during drought typically depends on political negotiations, alternative demand management and supply augmentation measures available to different entities, and other factors in addition to the water rights permit system.

Assigning water right priorities to maintaining reservoir storage levels relative to diversion rights is another issue. Protecting reservoir inflows is crucial to providing a dependable water supply. Each drawdown could potentially be the beginning of a several-year drawdown that empties the reservoir. However, forcing appropriators, with rights junior to the rights of the reservoir owner, to curtail diversions to maintain inflows to an almost full, or even an almost empty, reservoir is difficult and often is not the optimal use of the water resource. If junior diversions are not curtailed, the reservoir will likely later refill anyway, without any supply shortages occurring. Handling of the storage aspect of water rights is not yet precisely defined in Texas except for the Lower Rio Grande. The Lower Rio Grande is simpler in this regard because essentially all of the water users are supplied by two large reservoirs. Elsewhere, numerous reservoirs are owned and operated by various entities in the same river basin.

Although some recently issued permits specify the amount of the diversion to be returned to the stream, most permits do not. Return flows can significantly impact the availability of water to downstream users. This issue is currently being addressed particularly as related to programs to encourage reuse of wastewater effluents.

From the perspective of hydrology and water resources management, groundwater and surface water are closely interrelated. Use of one resource often has significant impacts on the other. However, there is only limited governmental control over the use of groundwater in Texas. Consequently, conjunctive management of ground and surface water resources is difficult. Depleting groundwater reserves are forcing a shift toward greater groundwater regulation.

Water availability depends upon water quality as well as quantity. The water supply capabilities of several major river systems that include very large reservoirs in Texas and neighboring states are severely constrained by salinity originating from natural salt deposits in geologic formations underlying the upper watersheds of the rivers. Salinity simulation capabilities have been added to WRAP for assessing the impacts of salinity on water availability [37]. Salinity problems are addressed in the regional plans [24]. However, incorporation of salinity into the water allocation systems is a complex issue yet to be resolved.

The Texas Water Code has required consideration of environmental flow needs in the water rights permitting process since 1985. Such needs include maintenance of aquatic habitat and species, water quality, public recreation, wetlands, and freshwater inflows to bays and estuaries. Although such needs have been considered in issuing permits since 1985, most water rights in the state have been granted without specifying instream flow requirements. Under mandates enacted by the Texas Legislature in 2001 and 2007, developing methodologies for establishing environmental flow criteria, establishing requirements for each river reach in the state, and incorporating them into the water rights system is currently a major focus.

6. Summary and conclusions

As water demands increase with population and economic growth, water allocation systems become increasing important worldwide. This chapter illustrates mechanisms for allocating water and highlights issues that must be addressed in their implementation. The Texas experience illustrates fundamental principles, issues, strategies, and complexities involved in developing and administering water allocation systems. Allocation of water resources among nations, states, regions, types of use, and numerous water users is a governing concern in water management in Texas as well as throughout the world.

Water allocation practices in Texas have evolved historically over centuries, with significant improvements occurring in recent years that continue to be refined. Texas shares water with Mexico and several neighboring states in the United States. Thousands of government agencies, cities, private companies, and citizens within Texas hold rights to use the waters of the state. Legal rights to use surface water differ from those for ground water. Surface water allocation for the Lower Rio Grande is different than for the rest of the state. With growing demands on limited water resources, expanding and refining water allocation systems has become a central governing concern in water management. The following observations regarding development and administration of water allocation systems in Texas are generally applicable in various other regions throughout the world.

- Water allocation systems include hierarchical systems of treaties and compacts between nations and states, ground water rights systems, surface water right systems, reservoir project ownership and operating practices, and contracts and agreements among a myriad of water management entities and water users. These water allocation mechanisms are overlapping and interconnected. Water allocation is integrally connected with other water resources planning and management activities.

- Water allocation practices evolve historically. In Texas, with growing demands on limited water resources, by the 1950's a disorganized myriad of water allocation practices had become a governing constraint to effective water management. Major change involving better water allocation mechanisms was necessary. The evolution of water allocation systems is continuing now and will continue in the future.

- Droughts motivate political action to improve water allocation systems. In Texas, major droughts led to milestone water management legislation being enacted in 1967 and 1997. The current drought that began in 2010 is expected to motivate additional legislative action that will contribute to the continued evolution of water allocation systems.

- The historical competition between agricultural, municipal, industrial, energy, and environmental water uses continues. There is a continuing shift from agricultural to municipal water use. Establishing and implementing environmental instream flow requirements is currently a major focus in Texas.

- Capabilities for assessing water availability and supply reliability within the framework of governing water allocation systems are essential for effective water resources planning and management. The Water Rights Analysis Package (WRAP) modeling system significantly contributes to effective water planning and allocation in Texas. WRAP is generalized for application to river/reservoir systems located anyplace in the world.

Author details

Ralph A. Wurbs

Texas A&M University, USA

References

[1] Santos Roman DM. Systematization of water allocation systems: an engineering approach. PhD dissertation, Texas A&M University, College Station; 2005.

[2] Rice L., White MD. Engineering Aspects of Water Law. Malabar: Krieger Publishing; 1991.

[3] Wallace J., Wouters P., editors. Hydrology and Water Law: Bridging the Gap. London: International Water Association (IWA) Publishing; 2006.

[4] Tofan C., Strambu S., editors. International Water Law, 2 volumes. Oisterwijk: ICA Press; 2008.

[5] Beach LB., Mammer J., Hewitt JJ., Kaufman E., Kurki A., Oppenheimer JA., Wolf AT. Transboundary Freshwater Dispute Resolution: Theory, Practice, and Annotated References. New York: United Nations University Press; 2000.

[6] Pricoli JD., Wolf AT. Managing and Transforming Water Conflicts. Cambridge : Cambridge University Press; 2009.

[7] Wolf AT., Natharius JA., Danielson JJ., Ward BS., Pender JK. International River Basins of the World, International Journal of Water Resources Development 1999; 14(4) 387-427.

[8] Grover VI., editor. Water: A Source of Conflict or Cooperation? Enfield: Science Publishers; 2007.

[9] Barlow M. The Global Water Crisis and the Coming Battle for the Right to Water. Toronto: McClelland & Stewart, 2007.

[10] Baillat A. International Trade in Water Rights: The Next Step. London: IWA Publishing; 2010.

[11] Roth D., Boelens R., Zwarteveen M., editors. Liquid Relations: Contested Water Rights and Legal Complexities. New Brunswick: Rutgers University Press; 2005.

[12] Selby J. Water, Power, and Politics in the Middle East. London: I.B. Tauris; 2003.

[13] Getches DH. Water Law in a Nutshell. 4th Edition. St. Paul: West Publishing, 2009.

[14] Beck RE., Kelley AK., editors. Waters and Water Rights, Cumulative Index, Volume 8, 2008 Edition. San Francisco: Matthew Bender (LexisNexis), 2008.

[15] Draper, SE., editor. Model Water Sharing Agreements for the Twenty-First Century. Reston: American Society of Civil Engineers, 2002.

[16] Eherat JW., editor. Riparian Water Regulations: Guidelines for Withdrawal Limitations and Permit Trading. Reston: American Society of Civil Engineers; 2002.

[17] Dellapenna JW. Appropriative Rights Model Water Code, Reston: American Society of Civil Engineers; 2007.

[18] Fischer D. The Law and Governance of Water Resources. Cheltenham: Edward Elgar Publishing; 2009.

[19] Hu D. Water Rights: An International and Comparative Study. London: IWA Publishing; 2006.

[20] Kissling-Nag I., Kuks S., editors. The Evolution of National Water Regimes in Europe: Transitions in Water Rights and Water Policies. London: Kluwer Publishers; 2004.

[21] van Koppen B., Giordano M., Buttersworth J. Community-based Water Law and Water Resource Management Reform in Developing Countries. Oxfordshire: CAB International; 2007.

[22] Longo PJ. Water Across Borders: Judicial Realities. In: Longo PJ., Yoskowitz DW. Eds. Water on the Great Plains: Issues and Policies. Lubbock: Texas Tech University Press; 2002.

[23] Water Resources Planning: Manual of Water Supply Practices, Manual M50. Denver: American Water Works Association; 2001.

[24] Water for Texas 2012. Austin: Texas Water Development Board; 2012.

[25] Texas Instream Flow Studies: Technical Overview, Report 369. Austin: Texas Commission on Environmental Quality, Texas Parks and Wildlife Department, Texas Water Development Board; 2011.

[26] Eaton DJ., Hurlbut D. Challenges in the Binational Management of Water Resources in the Rio Grande/Rio Bravo. U.S.-Mexican Policy Report No. 2. Austin: Lyndon B. Johnson School of Public Affairs, University of Texas; 1992.

[27] Hall GE. High and Dry: the Texas-New Mexico Struggle for the Pecos River. Albuquerque: University of New Mexico Press; 2002.

[28] Wurbs RA. Texas Water Availability Modeling System. Journal of Water Resources Planning and Management. American Society of Civil Engineers. 2005;131(4) 270-279.

[29] Wurbs RA. Water Rights Analysis Package (WRAP) Modeling System Reference Manual. 9th Edition. TR-255. College Station: Texas Water Resources Institute; August 2012.

[30] Wurbs RA. Water Rights Analysis Package (WRAP) Modeling System Users Manual. 9th Edition. TR-256. College Station: Texas Water Resources Institute; August 2012.

[31] Wurbs RA., Hoffpauir RJ. Water Rights Analysis Package Daily Modeling System. TR-430, College Station: Texas Water Resources Institute; August 2012.

[32] Wurbs RA. Fundamentals of Water Availability Modeling with WRAP. 6th Edition. TR-283, College Station: Texas Water Resources Institute; September 2011.

[33] Wurbs RA. Salinity Simulation with WRAP. TR-317, College Station: Texas Water Resources Institute; July 2009.

[34] Wurbs RA. Chapter 24 Water Rights Analysis Package (WRAP) Modeling System. In: Singh VP., Frevert D.K. (eds.) Watershed Models. Baton Raton: CRC Press; 2006.

[35] Wurbs RA. Chapter 1 Generalized Models of River System Development and Management. In: Uhlig U. (ed.) Current Issues of Water Management. Rijeka: InTech 2011.

[36] Wurbs RA., Schnier ST., Olmos HE. Short-Term Reservoir Storage Frequency Relationships. Journal of Water Resources Planning and Management. American Society of Civil Engineers. 2012;138(6).

[37] Wurbs RA., Lee and CH. Salinity in Water Availability Modeling. Journal of Hydrology. Elsevier Science, 2011;407(2) 451-459.